MEDDELANDEN FRÅN LUNDS UNIVERSITETS
GEOGRAFISKA INSTITUTIONER
avhandlingar 102

THE ROYAL UNIVERSITY OF LUND, SWEDEN

MEDDELANDEN FRÅN LUNDS UNIVERSITETS
GEOGRAFISKA INSTITUTIONER
avhandlingar 102

Remote Sensing of Cultivated Lands in the South of Sweden

Karin Hall-Könyves

Lund University Press

Lund University Press
Box 141
S-221 00 Lund,
Sweden

Art nr 20080
ISBN 91-7966-027-4

Chartwell-Bratt Ltd
Old Orchard Road
Bromley, Kent BR1 2NE
England

ISBN 0-86238-168-1

Printed in Sweden
Wallin & Dalholm Boktr AB, Lund

CONTENTS

ABSTRACT

The aim of the study is to investigate the potential for remote sensing application in Swedish agriculture with special reference to satellite-based monitoring.

The first part of the thesis describes preprocessing approaches including relative calibration, examination of various sun angle and atmospheric correction methods, and a study of the topographic effect. Two of the examined solar angle correction methods are concluded to be particularly suitable. The between-scene variability, which appears to be caused by atmospheric influence, is reduced by haze removal and standardization methods. The topographic conditions in the south of Sweden are found to have limited effects on LANDSAT data.

The second part of the work deals chiefly with crop classification and crop area estimation accuracy. In addition, the relationships between remotely-sensed data and crop types, crop varieties, crop yields and various amounts of insecticides and fertilization are studied. Low crop classification accuracies are generally noted, as in the case of cereals, whereas the crop area estimation accuracy for rape-seed and sugar beets is found to be high.

While acknowledging the existence of certain limitations, the study shows that remote sensing data is applicable on Swedish agricultural conditions. It indicates however, that further research is required to improve the methods used so far.

1

INTRODUCTION

Applications of remote sensing techniques in agriculture have attracted great attention in many countries for several years. In some of them remote sensing is incorporated in essential parts of agricultural research and management.

Some of the most interesting applications concern crop area estimation and assessment of crop status and crop production.

The first comprehensive agricultural inventory, called the Large Area Crop Inventory Experiment (LACIE) and undertaken in the years 1974 to 1977, was primarily initiated by the USA's interest to obtain information of the world's cereal production (Curran 1985). LACIE concerned estimation of the wheat production in the USSR, Latin America, China, Australia and was performed by multiplying the crop area derived from Landsat sensor data by the estimated crop yield derived from meteorological satellite data (MacDonald and Hall 1980). Each year this information was worth around $200 million to the USA's agricultural industry (Curran 1985). In 1980 LACIE was replaced by a monitoring programme named Agriculture and Resources Inventory Surveys Through Aerospace Remote Sensing (AgRISTARS). In this programme inventory of land use and assessment of land productivity constitute important parts. Recent studies by e.g. Ryerson et al. (1985), Moreira et al. (1986), McCloy et al. (1987) clearly demonstrate the capability of remote sensing for crop area estimation and an operational possibility is often indicated. Thus, with respect to crop areal estimates the remote sensing technique has proved to be a useful tool. Remote sensing data for assessment of crop status and crop production have however turned out to be unclear and the need for additional information sources is frequently underlined.

In Sweden remote sensing aerial and ground-based measurements of agricultural crops have been performed by e.g. Wastenson (1978), Kleman (1985), Nilsson (1986a,b,c).

Planning and control of Swedish agricultuural production are today associated with very high costs. The remote sensing technique is not frequently used in Swedish agriculture and an application could possibly provide increased effectiveness and accuracies and thereby create economical advantages.

The objecive of the present work was to increase the knowledge of the remote sensing potential for Swedish agricultural applications in general. For this purpose a project entitled "Studies of the influence of various environmental parameters on remotely-sensed data" was initiated during 1983 at the Remote Sensing Laboratory, University of Lund. The aim of the project was to achieve a better understanding of the remotely-sensed noise/target relationship with special references to agricultural monitoring. The following major studies were carried out to achieve the aim:

- A study of methods for reduction of between-scene variability (section 4).

- A study of the topographic effect in gently undulating terrain (section 5).

- A study of the relationship between remotely-sensed data and various crop parameters (sections 6 and 10).

- A study of the soil impact (section 7).

- A study of crop classification accuracies (section 7).

- A study of crop area estimation accuracies (sections 8 and 9).

The investigation included LANDSAT satellite data and to only a minor extent other remote sensing platforms. The report consists of two major parts. The first part (sections 3-5) deals with remote sensing preprocessing approaches. The second part (sections 6-10) elucidates remote sensing agricultural applications.

The work has been sponsored by the Swedish Board for Space Activities (DFR) and the Agricultural Engineering Department at the Swedish Sugar Beet Company.

For the accomplishment of this thesis I am indebted to many people who have assisted and encouraged me during the work. Not all of them can be mentioned here.

I am, however, especially grateful to my supervisor, Ulf Hellden, for his continuous support, to Eva Ahlcrona for long fruitful discussions, to Lennart Olsson who inspired me during the initial parts of the work, and to our guest researcher, Lu Jiaju, who has given me valuable help in several steps of the work.

I would also like to thank my family for their patience and generous understanding during the preparation of this thesis.

Finally I want to thank my competent secretary M. Vax and his likewise highly capable co-worker Allan for their efficient but, at times, somewhat capricious assistance.

2

STUDY AREA, DATA SOURCES AND EQUIPMENT

2.1 STUDY AREA

The study was performed in two regions in Sweden, the province of Scania and the province of Östergötland (Fig. 2.1). According to the Köppen classification, the province of Scania is dominated by a moist, temperate climate, with a rather warm summer. During the latest normal period (1931-60) the annual mean precipitation in the province was 633mm and the annual average temperature was +7.8°C (Loman 1986). The potential annual evaporation is approximately 550mm (Ångström 1974). The mean temperature is somewhat lower in the province of Östergötland. According to the soil map of Sweden (Troedsson and Wiberg 1986) the study was carried out in areas dominated by Eutric Cambisols and Dystric Cambisols (the FAO/ UNESCO soil classification system).

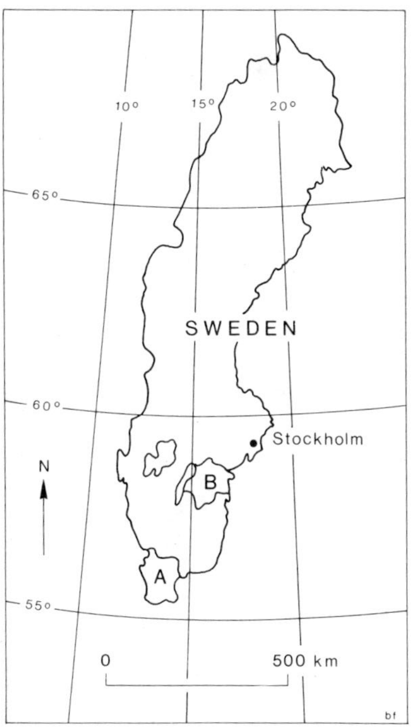

Figure 2.1 Location of the study areas.
A= the province of Scania,
B= the province of Östergötland.

2.2 DATA SOURCES

Remote sensing data included nine LANDSAT satellite acquisitions and one aerial photograph;

- LANDSAT data

Scene	Date	Satellite/ sensor	Solar elevation
194/21	May 19 1983	LANDSAT 4 MSS	50°
194/21	June 20 1983	LANDSAT 4 MSS	53°
194/21	July 06 1983	LANDSAT 4 MSS	52°
193/21	Sept. 01 1983	LANDSAT 4 MSS	38°
194/21	April 27 1984	LANDSAT 5 TM	44°
193/19 Q3	June 07 1984	LANDSAT 5 TM	51°
193/19 Q3	July 09 1984	LANDSAT 5 TM	50°
193/19 Q3	Aug. 26 1984	LANDSAT 5 TM	38°
194/21 Q4	Sept. 24 1986	LANDSAT 5 TM	30°

- Aerial photograph

  June 30 1986, altitude 600m, Kodak Aerochcrome Infrared Film 2443.

Ground data were obtained through:

- personal contact with individual farmers
- cooperation with the Central Bureau of Statistics (SCB)
- cooperation with the Swedish Sugar Company.

Additional data were received from topographical maps, soil maps, meteorological records, official publications by SCB and field controls.

2.3 LABORATORY RESOURCES

Several facilities were used in data processing and analyses:

- DIPIX ARIES II - advanced image processing system comprising tape station, disk memory (200 MB), video refresh memory, high resolution colour monitor, graphical 1 MB video memory, terminal and matrix printer. Host computer is a PDP 11/73.

- DIPIX ARIES III - includes tape drivers, 1 200 MB disk memory, high resolution colour monitor, 4 MB video memory and pixel processor. Host computer is a MicroVax II.

  The DIPIX systems contain extended software e.g. for geometric correction, radiometric correction, classification, filtering, principal component analysis etc. (ARIES is a trademark for DIPIX systems limited).

- EBBA I - an interactive, easily handled image processing system. The system is built on a Metric-85 micro computer (CPU 62 KB) and comprises tape station, video memory, colour monitor and TV-camera.

- Hi-state precision coordinate digitizer - the digitizing table has a 90cm X 120cm digitizing area, the accuracy is 0.127mm and the resolution is 0.02mm. The table is connected to the DIPIX ARIES systems.

- MicroVax II - the microvax system has 10 MB working memory and 1 200 MB disk memory.

- DIAPAC - originally the "Digital Image Analysis Package" was applied to the Sperry 1100/80 mainframe computer at the Lund University Computing Centre (Olsson and Pinzke 1984). The package is now available at the Microvax II system. DIAPAC contains conventional image analysis routines, texture transformations, spatial models, routines for integration of vector with raster data and image plotting. In the present work, DIAPAC was mainly used for generation of digital elevation models.

- MINITAB - a general purpose statistical system available at the MicroVax II system. MINITAB emphasizes aspects of statistics that are particularly appropriate for computer use (Ryan et al. 1985).

3

RADIOMETRIC AND GEOMETRIC CORRECTION

Accurate interpretation of remotely-sensed data requires radiometric and geometric correction prior to analyses. This pre-processing step is considered to be one of the basic tasks in image analyses.

3.1 RADIOMETRIC CORRECTION

According to Teillet (1986), two main groups of radiometric effects can be identified:

1. sensor-related effects

2. scene-related effects

3.1.1 The need for correction of sensor-related effects

Radiometric calibration is defined by two main procedures (Teillet 1986):

(i) absolute calibration (nominal calibration) for radiometric detector response

(ii) relative calibration

The radiance values (Appendix I) from the internal calibration system of the satellite and their corresponding voltage output from each detector are measured prior to launch of the satellite. In orbit, the internal calibration system and the sun are used to maintain the calibration between the radiance and the signal (Landsat Data Users Handbook 1984, Slater et al. 1987). However, according to Nelson (1985a), there are indications of uncalibrated sensor-changes with time. The absolute calibration of the detectors (raw digital numbers (DN) to calibrated DN) is normally performed at the receiving stations. All of the images included in this work were corrected according to absolute radiometric calibration.

Relative calibration accounts for detector-detector, band-band and time-time variations and is thus not only a prerequisite for multitemporal or multisensor studies but also if band ratioing is performed (this is neglected in several studies). The physical values that can be derived from the CCT digital numbers for each band are (i) radiance and (ii) reflectance (Appendix I).

Usually the radiance is calculated as in-band radiance (mW/(cm^2 sr)) or spectral radiance (mW/(cm^2 sr μm)) (Appendix I), (Robinove 1982, Nelson 1985a, Markham and

Barker 1986, Musick 1986, Markham and Barker 1987). In-band radiance is proportional to the bandwidth of the sensor while spectral radiance is related to sensor bandwidth only through the wavelength dependence of the observed radiation. The latter is therefore recommended by Price (1987).

In raw LANDSAT Multispectral Scanner and Thematic Mapper data and in the associated calibration data line dependent variations have been found. The variation that is caused by random fluctuations in the background reference level may be up to four digital counts (striping). This effect is especially significant in the widest bands. A radiometric de-striping technique is described by Murphy et al. (1984)

Examples of other error sources which may generate radiometric disturbances in especially TM data are:

- a scan angle effect; causes an increase in radiance towards the western edge of the scene (Malila et al. 1984, referred to by Fusco et al. 1986)
- a droop; the droop results in a systematic difference between cross-scan averages of forward and reverse scan, which is positive at the western edge and negative at the eastern edge of the scene (Malila et al. 1984, referred to by Fusco et al. 1986)
- Bright Target Saturation (BTS); a within-line sample location-dependent noise which is due to oversaturation after exposure of very bright targets (Barker 1984, referred to by Fusco et al. 1986). BTS includes a long memory effect and may therefore influence several quarter scenes.

In LANDSAT-4 MSS data a coherent noise pattern has been found (Tilton et al. 1985). This effect seems to have a spatial frequency of 3.6 pixels/cycle and a magnitude around 0.6 counts (Alford and Imhoff 1984, Likens and Wrigley 1985, Rice and Malila 1985 (referred to by Tilton et al. 1985), Malila 1985). Other radiometric aspects are modulation transfer function and quantization errors (Teillet 1986).

3.1.2 The need for correction of scene-related effects

3.1.2.1 Correction for sun angle and illumination variation

The solar elevation influences:

(i) the <u>amount</u> of airmass penetrated by the solar radiation

(ii) the solar illumination <u>angle</u> related to the scene.

The atmosphere has various subduing effects within various wavelengths and as the airmass is penetrated, the spectral distribution of the solar radiation is changed. The balance of solar irradiation between different LANDSAT bands is thus dependent on the optical airmass (m), which is dependent on the solar elevations and can be calculated as (Lindell et al. 1986)

$$m=1/\sin(h) \qquad (3.1)$$

where h= the sun elevation angle.

The radiance and the reflectance from targets are dependent upon the solar angle related to the scene in a very complex way that is mainly associated with (i) view angle, (ii) surface geometry, (iii) surface topography and (iv) atmospheric conditions. The relationship between these parameters has been studied by for example Kimes et al. (1980), Kollenkark et al. (1982), Kimes et al. (1984), Asrar et al. (1985a), Pinter et al. (1985), Ranson et al. (1985), Royer et al. (1985), Shibayama and Wiegand (1985). Curran (1985) recorded at a solar elevation of 31°, an altitude of ten meters, an 80% grass cover at various viewing angles. Curran (1985) found that "a move of only 10° from the vertical is seen to increase the difference in bidirectional reflectance and the red and near infrared wavebands from 28% to 49% with no change in LAI (Leaf Area Index) of the canopy or the hemispherical reflectance of the leaves" (Appendix I).

Since there is a considerable variation of the sun angle effect between various view angle/surface geometry/ surface topography/atmospheric condition combinations, this should be accounted for when applying solar elevation correction models. Concerning satellite applications, however, a common approximation is to assume Lambertian (Appendix I) surfaces (in spite of large parts of the Earth surface being non-Lambertian). A correction for illumination and geometric structure are more important for satellites with wide scan angles and with off-nadir viewing geometry, e.g. SPOT (27° off-vertical), LANDSAT Thematic Mapper (9° off-vertical) and AVHRR compared with LANDSAT Multispectral Scanner.

Smith et al. (1980) performed LANDSAT satellite analysis of terrain geometric effects on the optical scattering properties of Pinus ponderosa forests. The authors evaluated the Lambertian assumption for LANDSAT and examined various correction methods. On what conditions the Lambertian assumption may be correct was also studied. A photometric function developed by Minnaert (1941) was used to account for surface roughness. It was concluded that the Lambertian assumption for ponderosa pine is most valid for slopes of less than 25° and illumination angles of less than 45°.

Non-Lambertian methods to correct for solar angle effects are described by e.g. MacFarlane and Robinson (1984), Nelson

(1985a), Lindell et al. (1986), Markham and Barker (1986). These models include simple division by sun angle, calculation of spectral reflectance , ERIM transformations and consideration of the atmospheric subduing effect.

3.1.2.2 Correction for atmospheric effects

The radiance reaching a satellite sensor consists of:

(i) atmospheric radiance

(ii) background radiance

(iii) radiance from the target.

The atmospheric and background radiance are frequently referred to as path radiance. However, according to Deschamps et al. (1983) this term often is confused with atmospheric radiance and should therefore be avoided.

As the solar radiation passes through the atmosphere its spectral distribution, intensity, speed, frequency and direction are changed. This is due to (i) atmospheric scattering, (ii) absorption and (iii) refraction. These parameters are most active in visible and infrared wavelengths.

Atmospheric scattering mainly influences the direction of visible radiation and the spectral distribution of visible and near visible wavelengths. The two most important types of scattering are Rayleigh (causes haze in short visible wavelengths) and Mie scatter (Appendix I). Absorption influences wavelengths that are both shorter and longer than the visible wavelengths. Water vapour, carbon dioxide and ozone are the most common absorbers. When the atmosphere is turbulent (as it often is), refraction occurs as the radiation passes from one medium to another. This affects the geometric accuracy of the remotely sensed images (Curran 1985).

Effects due to atmospheric scattering and absorption may vary strongly both temporally and spatially. The sum of atmospheric scattering and absorption is called the atmospheric optical thickness (τ). The absolute atmospheric transmittance ($T\Theta = \exp(-\tau \sec\Theta)$) is impossible to measure in the absence of relevant meteorological and atmospherical data.

One of the most difficult variables to calculate is aerosol scattering. Accurate values may only be obtained by e.g. meteorological sounding balloons. Sometimes, the visibility is used as an approximation of aerosol concentration (Forster 1984). The ozone distribution and thickness are other components that are very complicated to measure.

Aranuvachapun (1986), Dalu (1986) and Ho et al. (1986) used data from meteorological studies to estimate the aerosol optical depth and the atmospheric absorption.

If a complete atmospheric correction is to be performed, the average background radiance must be determined. Studies by Dana (1982) showed that the radiance of an object is significantly affected by reflectance of background objects up to 500m away. Two methods to estimate background reflectance of a target are discussed by Forster (1984). These methods are either based on ground measurements of each cover type surrounding the target or on the known reflectance of one low and one high reflectance surface (Turner and Spencer 1972, Turner 1975, Kawata et al. 1978).

Satellite data, thus, may not undergo absolute atmospheric correction (i) without accurate meteorological- and atmospherical data and (ii) without knowledge of the background reflectance.

An example of a first-order atmospheric correction is the haze removal method (Sabins 1978, Robinove et al. 1981, MacFarlane and Robinson 1984, Curran 1985, Teillet 1986, Lillesand and Kiefer 1987). This method can only be regarded as a first step toward a complete atmospheric correction since it neglects the background radiance and the atmospheric absorption. It assumes that the atmospheric scattering can be estimated over complete black uniform surfaces, e.g. areas in shadow or clear deep water and that the atmospheric scattering will cause brightness values greater than zero for complete black areas. For each waveband these offset values are considered to be a function of the atmospheric scatter and are therefore subtracted from all pixel values within each band.

Another example of a first-order atmospheric correction of multitemporal images is the "standardization" approach discussed by Scarpace et al. (1978) and Lo et al. (1986). These authors used reflectance values from unchanging surfaces (airport runways and clear lake water) from each waveband at different dates. The image producing the smallest values was used as a standard and the other images were corrected to this image.

Estimations and studies of the atmospheric effect on remotely sensed data have been performed by several authors (Dave 1980, Jackson et al. 1983, Carnahan et al. 1984, Duggin et al. 1984, Kaufman and Fraser 1984, Spanner et al. 1984, Steven et al. 1984, Kaufman 1985, Lord et al. 1985b). Various correction approaches have been studied by e.g. Crist (1984), Forster (1984), Kaufman and Fraser (1984), MacFarlane and Robinson (1984), Spanner et al. (1984), Kaufman (1985), Lindell et al. (1986), Lo et al. (1986), Singh and Cracknell (1986), Steven and Rollin (1986).

Steven and Rollin (1986) point out that it is particularly important to perform atmospheric corrections when measuring physical parameters on the surface. These measurements may include multitemporal studies of crops and various crop growth estimations. According to these authors, studies relating to differences between classes (classification) may be successful without atmospheric corrections. The necessity of regarding the atmospheric effect in classification procedures as well was elucidated by Kaufman and Fraser (1984) and Kaufman (1985) maintaining that:

- The atmospheric effect creates differences between the reference and the test field in the classification processes.

- The atmospheric effect reduces the separability of field classes. This is caused by a broadening of the radiance distribution of a given class due to nonuniform contamination of the upward radiance above a given field by light reflected from surrounding fields, and then scattered into the instantaneous field of view (IFOV, the field of view of a detector at one instrument) by the atmospheric constituents. The result is a larger overlap between various classes that obstructs the classification possibilities.

3.1.2.3 Topographic consideration

Measured spectral radiances may vary according to slope and aspect characteristics (Teillet 1986) and many researchers in remote sensing have demonstrated that terrain topography may influence remotely-sensed data (Justice 1978, Strahler et al. 1978, Holben and Justice 1981, Stohr and West 1985)

3.2 GEOMETRIC CORRECTION

LANDSAT imagery that has been corrected for e.g. Earth curvature and rotation and satellite altitude errors still contains geometric distortion. For some applications these errors are unimportant but for several investigations (multitemporal studies and direct comparisons with maps) geometrically transformed imagery is needed. This transformation includes selection of a number of ground control points, which are identifiable on both the image and a map. The transformation matrix between image and map co-ordinates is determined and using interpolation technique the scene is resampled (Davison 1986).

3.3 PREPROCESSING APPROACH

3.3.1 Radiometric correction

3.3.1.1 Calculation of spectral radiance

Conversion from digital numbers on CCT to scientifically meaningful units such as spectral radiance ($L\lambda$) was accomplished through the following equation (Markham and Barker 1986).

$$L_\lambda = Lmin_\lambda + \frac{Lmax_\lambda - Lmin_\lambda}{DNmax} \times DN \qquad (3.2)$$

where

DN = Digital numbers on CCT

DN_{max} = Range of rescaled radiance in DN

$L_{min\lambda}$ = Minimum spectral radiance (mW/cm^2 sr μm) (from Markham and Barker 1986)

$L_{max\lambda}$ = Maximum spectral radiance (mW/cm^2 sr μm) (from Markham and Barker 1986)

L_λ = Spectral radiance (mW/cm^2 sr μm)

Table 3.1 Post-calibration dynamic ranges and mean solar exoatmospheric spectral irradiances for the LANDSAT satellites used. Valid for U.S. and Esrange processed data. If a conversion from spectral radiance to in-band radiance is required, multiply the spectral radiance with the bandwidth.

	Bandwidth (μm)	Lminλ	Lmaxλ	DNmax	Esunλ
Landsat 4 (> April 1 1983)					
MSS 1	0.5-0.6	0.4	23.8	255*	185.1
MSS 2	0.6-0.7	0.4	16.4	255*	159.3
MSS 3	0.7-0.8	0.5	14.2	255*	126.0
MSS 4	0.8-1.1	0.4	11.6	255*	87.8
Landsat 5					
TM 1	0.45-0.52	-0.15	15.21	255	195.7
TM 2	0.52-0.60	-0.28	29.68	255	182.9
TM 3	0.63-0.69	-0.12	20.43	255	155.7
TM 4	0.76-0.90	-0.15	20.62	255	104.7
TM 5	1.55-1.75	-0.037	2.719	255	21.93
TM 7	2.08-2.35	-0.015	1.438	255	7.452

Spectral radiances, Lminλ and Lmaxλ, in mw/(cm^2 ster μm), Esunλ, in mW/(cm^2 μm).
*= normally 127 DN but for the present data 255 DN (Esrange data)

Tab. 3.1 provides the band-specific parameters that have been used.

For the MSS data included in the sun angle correction method no. 2, presented below, the in-band radiance and not the spectral radiance was calculated.

Arithmetic transformations were performed on a DIPIX ARIES III system.

The author considers it to be beyond the goal of this work to perform analyses and corrections of all sensor-related and scene-related radiometric effects discussed in section 3.1.1. These effects may however certainly influence results of image analyses and image interpretation and should therefore be added to study limitations.

3.3.1.2 Correction of sun angle and irradiance variation effects

Five methods for correction of sun angle and irradiance (Appendix I) variation effects were used.

1. Normalization for solar irradiance by transforming spectral radiance calculated in equation 3.2 to approximate effective at-satellite reflectance (Markham and Barker 1985, 1986).

$$\rho_{p\lambda} = \frac{\pi \times L_\lambda \times d^2}{Esun_\lambda \times \cos\theta_s} \qquad (3.3)$$

where

$\rho_{p\lambda}$ = Unitless effective at-satellite reflectance

L_λ = Spectral radiance at sensor aperture in (mW/(cm² sr μm)) from equation 3.2

d = Earth-sun distance in astronomical units from nautical handbook

$Esun_\lambda$ = Mean solar exoatmospheric spectral irradiances in (mW/(cm² μm)) (from Markham and Barker 1986).

θ_s = Solar zenith angle in degrees

Values of $Esun\lambda$ are presented in Tab. 3.1.

Thus, spectral radiance measurements were converted to spectral reflectance measurements by accounting for the strength of the exoatmospheric spectral irradiance (Appendix

I), the solar angle and the Earth/sun distance. Although annual changes in the Earth/sun distance may account for a large change in the solar constant, such a correction is in the noise level (systems and atmospheric noise are as large as changes that might be imparted to the data by the changing Earth/sun distance) and are probably of no practical utility. (Note that this method calculates reflectance from the target, the background and from the atmosphere).

2. Correction of the sun angle effect by using a sun angle of 90° as reference, corresponding to an optical airmass m=1 (Lindell et al. 1986).

$$B_{\lambda,s} = B_{\lambda}\tau^{1-m}{}_{\lambda} \qquad (3.4)$$

where

m = 1/sin(h), m= optical airmass, h= the sun elevation angle

$B_{\lambda,s}$ = Calculated solar angle corrected or normalized radiance

B_{λ} = In-band radiance

τ_{λ} = Atmospheric transmission constant

Atmospheric transmission data were obtained from Fors (1986) and Lindell et al. (1986).

3. Correction of the sun angle effect by dividing $L\lambda$ by the cosine of the solar zenith angle (MacFarlane and Robinson 1984).

$$L_{\lambda,c} = L_{\lambda}/\cos_{\theta s} \qquad (3.5)$$

where

L_{λ} = Spectral radiance, calculated from equation 3.2

θs = Solar zenith angle in degrees

$L_{\lambda,c}$ = Solar angle corrected radiance values

4. Normalization of all data to the same solar zenith angle (Spanner et al. 1984).

$$L_{\lambda,no} = L_{\lambda}\cos_{\theta(1)}/\cos_{\theta(2)} \qquad (3.6)$$

where

L_{λ} = Spectral radiance, calculated from equation 3.2

$\theta(1)$ = The solar zenith angle to normalize to

$\theta(2)$ = The solar zenith angle to be normalized

$L_{\lambda,no}$ = Normalized radiance values

5. Ratio (MSS4/2, TM4/3 of DN, L_{λ}, $L_{\lambda,c}$, $L_{\lambda,no}$, $\rho_{p\lambda}$ and $B_{\lambda,s}$.

Within different LANDSAT spectral bands the balance of solar irradiation reaching the ground changes with sun angle. Only method 2 accounts for this while methods 3-5 assume that the solar elevation effect is the same in all bands.

Methods 1-5 assume that the targets are Lambertian.

As has been discussed in section 3.1.2, the Lambertian assumption is not valid for the main part of natural surfaces. However, the author considers it to be outside the objective of this work to develop non-Lambertian satellite models and therefore the Lambertian approximation has been used.

3.3.1.3 Atmospheric correction and topographic consideration

Two atmospheric correction methods were used and compared:

1. haze removal

2. atmospheric standardization to reference surfaces

These two methods are discussed in section 4.

The topographic effect on LANDSAT MSS and TM data are presented by Hall-Könyves (1987) and in section 5.

3.3.2 Geometric correction

Radiometrically calibrated images were geometrically corrected by first performing a resampling of one image to topographical maps (1:50 000). All other images covering the same area were then resampled "image to image" to this. The resampling algorithm was a third degree polynom and cubic convolution, based on 30-50 ground control points within each topographical map sheet (25x25km). The residuals at the ground control points were around 0.5 pixels. A DIPIX ARIES image analysis system was used and through the resampling procedure pixel sizes were reduced to 50m (MSS pixels) and 20m or 25m (TM pixels).

4

A STUDY OF METHODS FOR REDUCTION OF BETWEEN-SCENE VARIABILITY

The spectral variability of Earth features over time is a function of:

(i) changes in the spectral response characteristics of targets

(ii) viewing conditions.

For the main part of remote sensing applications the first factor is of interest and contains useful information while the second is considered to be noise. For multitemporal studies it is thus important to correct for the latter and consider changes which are due to:

(i) sensor differences

(ii) effects of changing sun angle and illumination

(iii) atmospheric conditions.

4.1 OBJECTIVES

The present work includes multitemporal studies and therefore changes due to sun angle and atmospheric conditions must be accounted for. The purpose of this section was to:

1. evaluate various transformations for reducing scene variability with emphasis on viewing factors

2. reduce scene variability with emphasis on possible atmospheric effects

4.2 REDUCTION OF BETWEEN-SCENE VARIABILITY WITH EMPHASIS ON VIEWING FACTORS.

Some commonly used transformations connected to correction of viewing factors were studied in order to decide which transformations were most effective. In addition, climatological data were used to study the relationship between the meteorological situation and the scene variability. Two multitemporal data sets were included:

(i) MSS data, the province of Scania (1983: May 19, June 20, July 06).

(ii) TM data, the province of Östergötland (1984: June 07, July 09, August 26).

4.2.1 Methods

The two main areas covered by MSS and TM scenes, respectively (presented in Fig. 2.1) were divided into a number of sub-areas (Fig. 4.1). For each sub-area the satellite response of several "unchanging" surfaces (surfaces that can be assumed not to change with time) was measured at various dates at a DIPIX ARIES system. The surfaces were measured according to the following transformations (section 3):

- DN and L_λ
- $L\lambda$,no
- $L\lambda$,c
- $\rho p\lambda$
- $B\lambda$,s
- ratioing (MSS4/2, TM4/3 of DN, $L\lambda$, $L\lambda$,no, $L\lambda$,c, $\rho p\lambda$ and $B\lambda$,s)

The following "unchanging" surfaces were used:

- sea surfaces
- lake surfaces
- cross-roads
- airport runways

The targets were chosen for the stability of their satellite response over time and efforts were made to avoid edge pixels. The approximate size of the studied "unchanging" surfaces is presented in Tab. 4.1.

To determine which transformation that most effectively reduces scene variability, the coefficient of variation (CV= standard deviation/mean) was calculated for each transformation/band/"unchanging" surface combination (CV was used so that scaling differences between the transformations could not affect the comparison). The transformation which yielded the lowest CV values would be the most effective preprocessing approach for correcting between-scene variations connected to viewing factors and system factors (Nelson 1985b). Thus, the lower the CV, the better the correction of reduction of between-scene variability.

Meteorological data associated with each acquisition were collected from various meteorological stations. Information concerning temperature (°C), precipitation (mm), wind speed (m/s), at-surface visual range (km), cloud cover (x/8), relative humidity (%) and atmospheric pressure (mbar) was thereby obtained. The location of the stations are shown in Fig. 4.1.

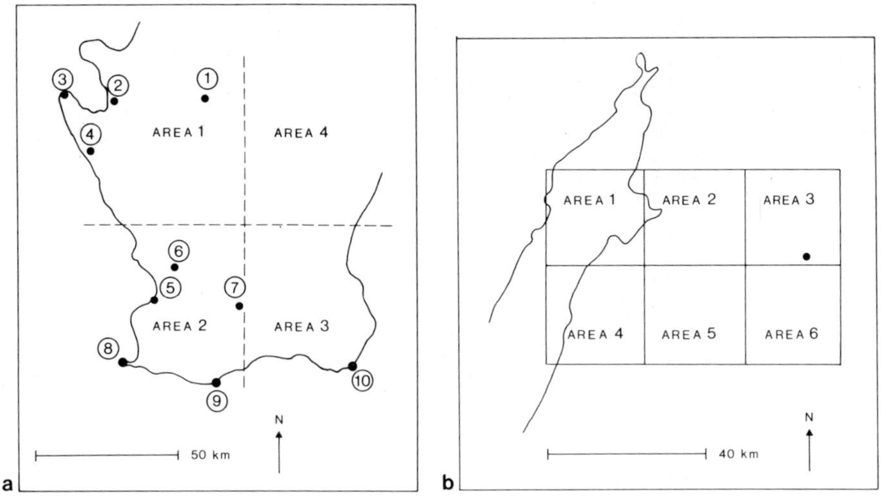

Figure 4.1 Sub-areas of (a) the province of Scania and (b) the province of Östergötland. Subareas 1-6 of Östergötland correspond to topographical map sheets 8ENO, 8FNV, 8FNO, 8ESO, 8FSV, 8FSO. ●= meteorological station.

Table 4.1 The approximate size(*) of the studied "unchanging" surfaces.

The province of Scania

	Area 1	Area 2	Area 3	Area 4
Sea surface	40 000	40 000	60 000	60 000
Lake surface	1 500	1 000	900	1 100
Cross-roads	10	8	-	-
Airport runways	15	13	-	-

The province of Ostergotland

	Area 1	Area 2	Area 3	Area 4	Area 5	Area 6
Lake surface	8 000	15 000	28 000	12 800	800	1 000
Cross-roads	30	20	30	50	20	20
Airport runways	-	-	300	-	-	-

(*)= number of pixels

4.2.2 Results

4.2.2.1 The between-scene variability

The coefficient of variation was calculated for each sub-area (Fig. 4.1) and for each transformation/band /"unchanging" surface combination.

The results implied that the magnitude of the CV is dependent on surface type (brightness) as well as on the transformation approach. Similar findings were obtained from all sub-areas of MSS as well as TM data, even though the absolute values of CV varied. The following was noted:

- the transformations consistently exhibiting the lowest CVs were $L\lambda,no$, $L\lambda,c$, $B\lambda,s$ $\rho p\lambda$ and their ratios
- for all transformations, dark surfaces (water) produced larger CVs compared with rather bright targets (e.g. cross-roads)
- for dark surfaces ratios generated high CVs, while for bright surfaces ratios produced low CVs

Ratioing is an effective variance reduction technique only if additive effects are small compared with multiplicative effects. The additive effects are relatively large over dark targets (Nelson 1985b) and this may account for the high CVs of the ratios.

The CVs of $L\lambda,no$, $L\lambda,c$, $B\lambda,s$ and $\rho p\lambda$ were equal as were the CVs of their ratios. The magnitudes of the CVs were approximately 5-25% and thus some target variation still remains. This "remaining" variations may be:

(i) in the noise level and due to factors discussed in section 4.3

(ii) caused by changes in the atmosphere.

4.2.2.2 The relationship between the scene variability and meteorological data.

According to Nelson (1985b), several meteorological parameters may explain between-scene variability. To examine if the "remaining" variation can be connected with the meteorological situation (i.e. to a certain extent the atmospheric condition) the relationship of between-scene variability and meteorological data was studied.

For each sub-area (Fig. 4.1) and "unchanging" surface, solar angle corrected data were plotted against acquisition date. Some examples of these plots are given in Figs. 4.2-4.3. The data presented in these figures are the mean of each "unchanging" surface in the sub-area.

For the MSS data the $L_{\lambda,no}$, $L_{\lambda,c}$, $B_{\lambda,s}$ and $\rho_{p\lambda}$ values varied for various dates (Fig. 4.2). Compared with the June 20 and July 06 acquisitions the May 19 and September 01 measurements yielded higher values of solar angle corrected data. Thus, July 06 produced the lowest values, then June 20, May 19 and finally September 01 exhibited the highest values.

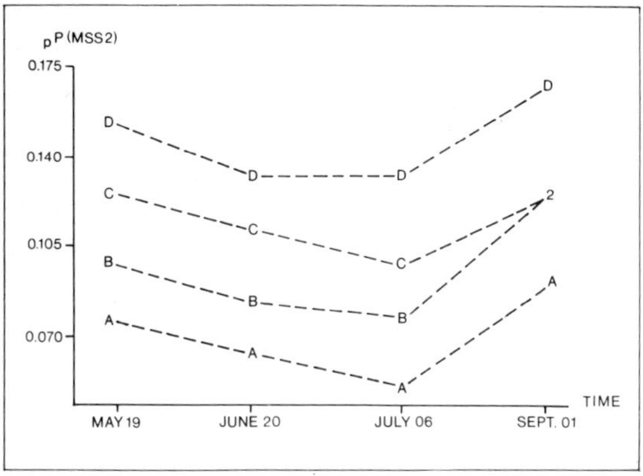

Figure 4.2 The at-satellite spectral reflectance (MSS2) of various targets, plotted against acquisition date (one sub-area). Target
A= sea surface,
B= lake water,
C= cross-roads,
D= airport runway.

This pattern is valid for all MSS band/sub-area/"unchanging" surface/$L_{\lambda,no}$, $L_{\lambda,c}$, $B_{\lambda,s}$ and $\rho_{p\lambda}$ combinations. The magnitude of the variation between the acquisition dates varies however with band (MSS2 higher than MSS4), with surface type (dark targets higher than bright targets) and with sub-area (sub-area 2 and possibly sub-area 1 higher than sub-areas 3 and 4).

For the acquisitions of the province of Östergötland differences between $L_{\lambda,no}$, $L_{\lambda,c}$, $B_{\lambda,s}$ and $\rho_{p\lambda}$ of various dates were obtained as well (Fig. 4.3). This variation was obvious in TM1-4 while TM5 and TM7 did not show any variation at all. For this data set the July 09 acquisition always yielded enhanced values compared with the June 07 and August 26 acquisitions (June 07 and August 26 were very similar). This was valid for all sub-area/"unchanging" surface/$L_{\lambda,no}$, $L_{\lambda,c}$, $B_{\lambda,s}$ and $\rho_{p\lambda}$ combinations (except sub-areas 4 and 6 which produced very high values on August 26 compared with June 07). As for the MSS data the magnitude of the variation was connected to wavelength and surface type.

The meteorological factors (i) at surface visual range (km), (ii) cloud cover (x/8), (iii) atmospheric pressure (mbar) and (iv) relative humidity (%) were for each meteorological station (Fig. 4.1) plotted against acquisition date. Some examples of these plots are given in Fig. 4.4.

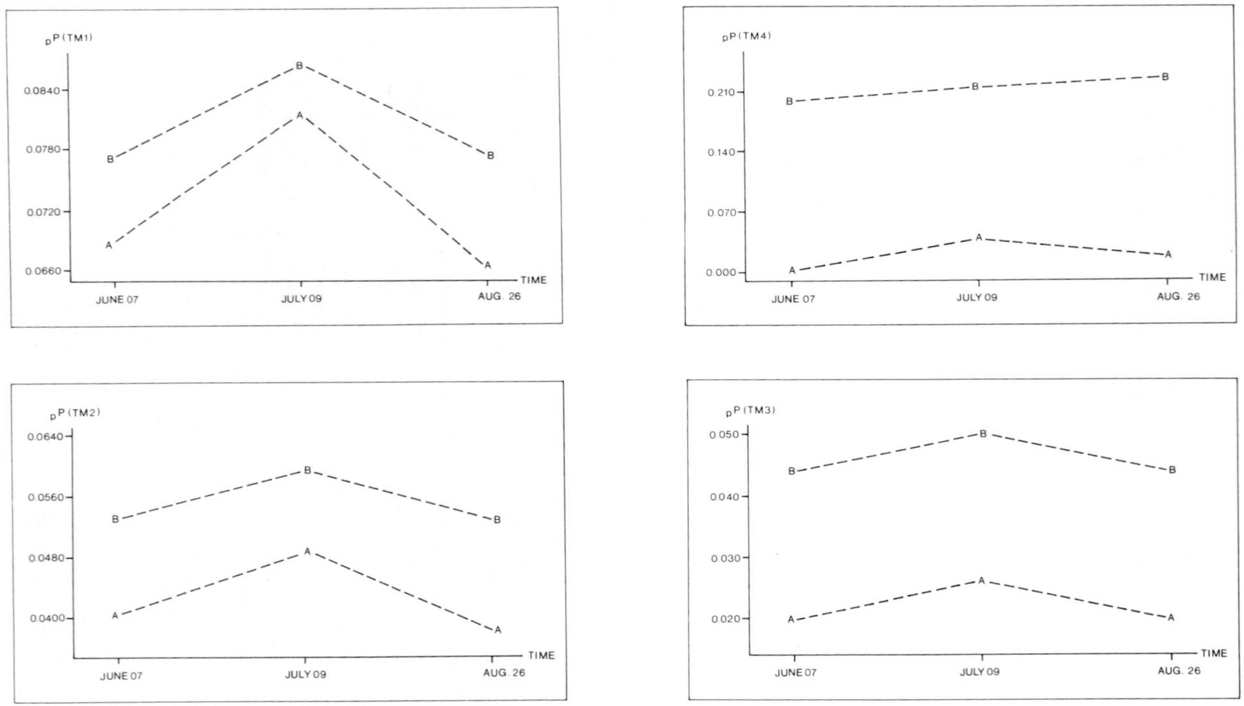

Figure 4.3 The at-satellite spectral reflectance (TM1-4) of two targets, plotted against acquisition date (one sub-area).
Target A= lake surface, B= cross-roads.

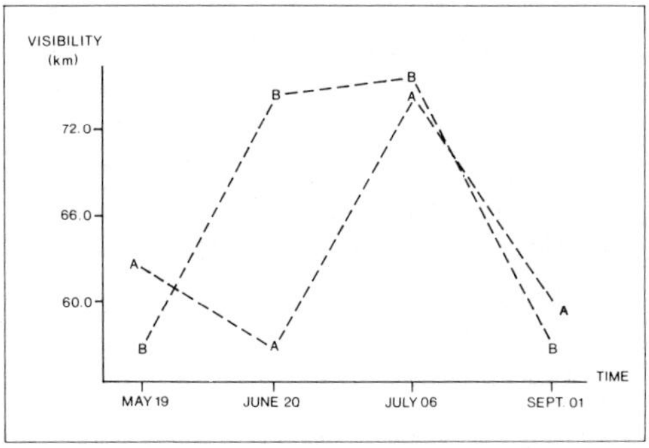

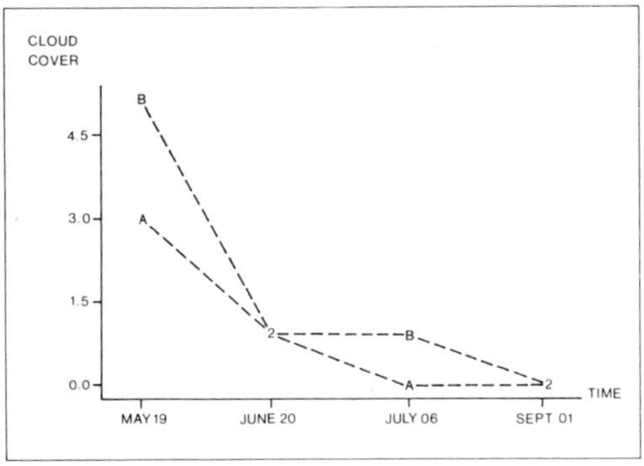

Figure 4.4 Visibility and cloud cover plotted against acquisition date (one sub-area, the province of Scania). A= at. 7 a.m., B= at 1 p.m.

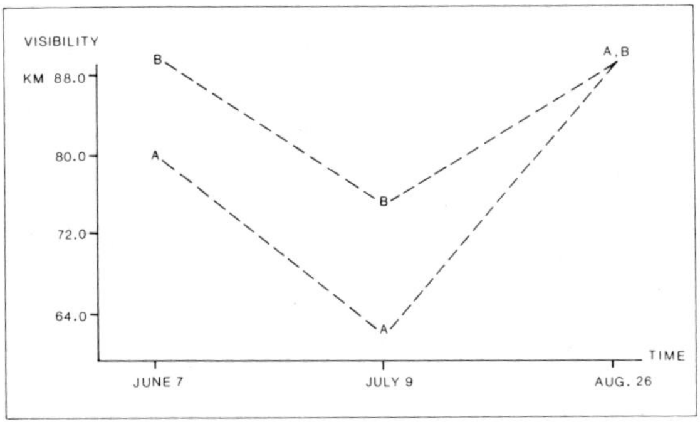

Figure 4.5 Visibility plotted against acquisition date (one sub-area, the province of Östergötland). A= at. 7 a.m., B= at 1 p.m.

Most of the meteorological stations in Scania revealed a common trend concerning the meteorological parameters (Fig. 4.4):

- the visibility (km) and the atmospheric pressure (mbar) were higher on the 20th of June and 6th of July than on the 19th of May and 1st of September

- the cloud cover (x/8) and the relative humidity (%) were lower on the 20th of June and 6th of July than on the 19th of May and 1st of September.

Data, available at the meteorological station in Östergötland were plotted against acquisition date as can be seen in Fig 4.5. These figures show that a low visual range was recorded on the 9th of July compared with the other two dates. An example of at-satellite reflectance ($\rho p\lambda$) at various visual ranges is presented in Fig. 4.6.

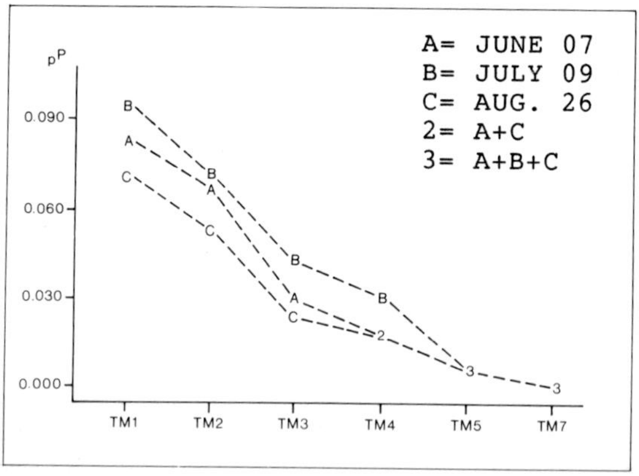

Figure 4.6 The at-satellite spectral reflectance (TM1-5,7) of lake water (the province of Östergötland, one sub-area). A,C= "comparatively" high visibiliy, B= "comparatively" low visibility.

When comparing Fig. 4.2 by Fig. 4.4 and Fig. 4.3 by Fig. 4.5 and when studying Fig. 4.6, correspondences were revealed. Thus, comparatively high values of sun angle corrected LANDSAT data were connected with low visibility and atmospheric pressure and with high cloud cover and relative humidity, while low values of sun angle corrected LANDSAT data were related to high visibility and atmospheric pressure and with low cloud cover and relative humidity. This was a general relationship, valid for the main part of the data set. A rather intimate association between sun angle corrected LANDSAT data and meteorological parameters was thus obtained.

Lo et al. (1986) maintained that between-scene variations (over time) of targets that are assumed not to change with time may be attributed to changes in atmospheric conditions. Relatively low radiance values would imply low atmospheric scattering and comparatively high radiance values would intimate high atmospheric scattering. According to Lo et al. (1986), it is thus possible to explain the obtained "remaining" variations by changes in atmospheric scattering. This, together with the observed close connection between sun angle corrected LANDSAT data and meteorological factors (which at least to a certain extent are dependent on atmospheric conditions) denotes that the used LANDSAT scenes may be disturbed by atmospheric effects.

Thus, the results of this section indicated that the "remaining" variation (section 4.2.2.1) may be caused by changes in the atmosphere. This was however not statistically proved and it cannot be excluded that the "remaining" variation (or some part of it) may be in the noise level or be due to factors discussed in section 4.4. Nevertheless, since rather strong indications of atmospheric disturbances were obtained, the author considers it unsatisfactory to neglect the "possible" atmospheric impact. Therefore, prior to further use of the LANDSAT data, efforts were made to account for atmospheric conditions.

4.3 REDUCTION OF BETWEEN-SCENE VARIABILITY WITH EMPHASIS ON ATMOSPHERIC INFLUENCE

4.3.1 Haze removal

The effect of sky light and haze is often suppressed through the simple "haze removal" or "darkest pixel subtraction". This method is based on the assumption that the longest wavelengths are unaffected by haze and that this wavelength can be used as a yard stick of the haze effect in the other wavelengths (Sabins 1978, Curran 1985). The "darkest pixel subtraction" is accomplished by performing a subtraction of the darkest pixel within each wavelength. As reference surfaces Sabins (1978) and Curran (1985) recommended areas in shadow. However, water is used in several studies. In the visible wavelengths, around 5% of the electromagnetic energy is reflected by the water (Curran 1985) and therefore the present author considers it unsuitable to use water as reference surfaces.

In this study relatively smooth terrain has been investigated and the presence of steep slopes yielding shadows so large that they are unaffected by background radiance are rare. Thus, for the present study the "darkest pixel subtraction" method is not suitable. However, as the method has been used in an extensive number of studies, it was applied also in the present study in spite of the fact that it may not be the best fitted.

In addition to the haze removal method a standardization approach was applied. These methods represent only first-order corrections. Relevant meteorological and atmospherical data, and values of background reflectance were not available, and therefore it was impossible to accomplish an absolute atmospheric correction.

4.3.2 Standardization of the scenes

Scenes, acquired at various dates, were standardized through a method suggested by Scarpace et al. (1978) and Lo et al. (1986). This method implies that the image yielding the smallest reflectance value (solar elevation corrected data) of an "unchanging" target, within a multitemporal data set, may be used as a standard. The standard is subtracted from the reflectance values of this target, observed in the same band on the other dates and the resulting remainder is considered to be atmospheric scattering in each image. These values are subtracted (for each band) from all the pixels of the respective images.

When applying this method to the present data set, some questions arose:

(i) Which solar elevation correction method should be used?

(ii) Should the division of the data set into sub-areas (Fig. 4.1) be utilized? Expressed differently, would it be possible to apply the standardization method on each sub-area and thereby obtain a standardization procedure valid not only for temporal differences but for spatial differences as well (each sub-area would in that case be changed differently according to its own variance)? The alternative is to use the "same" standardization value in the total image.

(iii) How many and which "unchangning" targets should be included? All "unchanging" targets are associated with problems and limitations (section 4.4). A standardization procedure based on only dark surfaces may "overcompensate" for bright surfaces (dark targets yield larger differences between scenes than bright targets). On the other hand, a procedure based on only bright targets may "undercompensate" for dark targets.

The following arguments were considered:

(i) Solar angle techniques like $L\lambda,no$ and $L\lambda,c$ consider only variation of the solar illumination angle related to the scene and thus neglect the amount of airmass penetrated. The at-satellite reflectance procedure ($\rho p\lambda$) however, considers the mean solar exoatmospheric spectral irradiance while the $B\lambda,s$ considers the optical air mass and the atmospheric transmission constant. Therefore the two latter were concluded to be more suitable than the other approaches.

(ii) Since the standardization method is not used very often, the accuracy of it is not well known. Therefore the method should not be exaggerated by performing a spatial standardization procedure until the accuracy of the method is better known.

(iii) Due to the uncertainty of the "unchanging" targets the standardization procedure should be based on the mean value of all "unchanging" targets.

Thus, the standardization procedure was performed according to the following:

- MSS data (the province of Scania).

- Since the July 06 acquisition produced the lowest values of $\rho p\lambda$ and $B\lambda,s$ this scene was chosen as a standard.

- For each band/$\rho p\lambda$ and $B\lambda,s$ combination, the "mean standard" was subtracted from each of the other scenes.

- The "minimum" remainder obtained for each image/ band/$\rho p\lambda$ and $B\lambda,s$ combination was noted.

- The "minimum" remaining values were subtracted from all the pixels in each scene.

- TM data (the province of Östergötland).

- The August 26 acquisition was chosen as a standard and according to the method, described above, the other scenes were standardized to this scene.

4.4 STUDY REMARKS

* Asphalted surfaces like cross-roads and airport runways represent rather small areas. Measurements of small targets are influenced by background radiance (Ranson et al. 1986). This means that the areas surrounding a small surface target influence the sensor response of this target. When comparing measurements of small "unchanging" surfaces from various dates, the possible difference may thus be due to variation of background radiance intensity.

* Moisture on asphalted surfaces may effect the satellite response of these surfaces.

* The use of measurement data for sea water and lakes are based on neglect of the light backscattered by water compared with the atmospheric haze brightness. In the case

of turbid water, this procedure is not relevant (Kondratyev et al. 1986).

* The atmospheric "correction" approaches are rather rough methods and they represent only a first-order correction for additive atmospheric effects. Possible effects of atmospheric absorption have not been accounted for.

* No spatial atmospheric correction was performed.

* All methods used for solar elevation effect correction assume Lambertian surfaces. Royer et al. (1985:1899) maintain that "to compare measurements performed under different geometric and illumination conditions, one must take into account atmospheric effects as well as the angular anisotropy of the reflective properties of natural surfaces".

4.5 SUMMARY

The purpose of this section was to evaluate various transformations for reduction of scene variability with emphasis on viewing factors (sun elevation effects) and to reduce possible atmospheric influences.

Two sun elevation correction methods ($\rho P\lambda$ and $B\lambda,s$) were concluded to be more suitable than the others. A close relationship between meteorological data and between-scene variability was obtained, which indicated that the scenes at least to a certain extent are influenced by atmospheric effects. The haze removal method was applied and a standardization procedure was developed to reduce between-scene variations caused by atmospheric effects.

5

THE TOPOGRAPHIC EFFECT ON LANDSAT DATA IN GENTLY UNDULATING TERRAIN

5.1 INTRODUCTION

The topographic effect is defined as the variation in response from inclined surfaces compared to response from a horizontal surface as a function of the orientation of the surface relative to the light source and sensor position (Holben and Justice 1981, Justice et al. 1981). For a uniform cover class the topographic effect can produce a wide range of pixel values and thus influence the accuracy of information extracted from remotely-sensed data.

Earlier investigations concerning the topographic influence have mainly been performed in fairly rugged terrain and predominantly by airborne and ground-based equipment. Only limited information concerning the possible topographic effect on LANDSAT satellite data in gently undulating areas is therefore available.

The aim of the presented study was to determine the relationship between various topographic parameters (angle of incidence, slope degree and slope aspect) and LANDSAT response variations (Hall-Könyves 1987).

5.2 REVIEW OF PREVIOUS FINDINGS

Many researchers in remote sensing have demonstrated that terrain topography may influence remotely sensed data (Justice 1978, Strahler et al. 1978, Holben and Justice 1980, Justice et al. 1981, Stohr and West 1985).

Using a hand-held radiometer Holben and Justice (1980) obtained a range of 50 pixel values (128 digital output levels) for a uniform sand surface. This was explained solely by variations in slope angle and aspect. The percentage change in response relative to the horizontal of, for example, slope magnitude 10° and aspect string 0°, 45°, 90°, 135° and 180° was 17.7, 7.6, 1.5, 9.1 and 15.5 per cent.

Justice et al. (1981) quantified and reduced the topographic effect on LANDSAT MSS imagery of a uniform forest cover with three major slope units (9°, 15°, 22°) at a solar elevation of 55°. Four methods for reducing the topographic effect were compared. The nort-west slope of the ridge was visibly darker than the south-east slope. This was most obvious for MSS3 and 4. Band ratioing did not eliminate the topographic effect. Of the three radiance models studied the Lambertian and modified Lambertian models increased the topographic effect, while the non-Lambertian model reduced the effect by 86 per cent.

Stohr and West (1985) examined the influence of various terrain elements (slope aspect and slope degree) and Sun-scanner look angle upon airborne multispectral scanner response of a single cover type, a pasture. The authors found that the terrain elements accounted for appreciable variation in response in all wavebands.

5.3 APPROACH

5.3.1 Study area and ground cover data collection

The study areas, located in Scania in the south of Sweden, cover 840km² (Fig. 5.1). 9400ha of this surface was studied. The steepest slope was approximately 60° and the mean gradient about 6-7°. The fertility of the soils is one of the highest in Sweden and the region is dominated by agricultural activities. The land use and vegetation cover of study areas A and B are characterized by cultivated fields and some pasture areas. The content of woodland is quite extensive in area B. Forest cover and pasture areas constitute the greater part of area C.

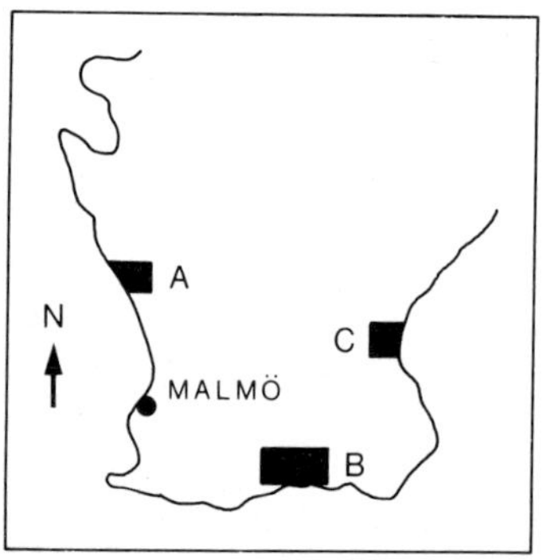

Figure 5.1 Scania in the south of Sweden: location of the three study areas. Study area A: x, 1310-1323; y, 6200-6208. B: x, 1360-1389; y, 6141-6160. C: x, 1388-1399; y, 6175-6188. (RAK-system coordinates).

Three major types of homogeneous cover were examined, namely cultivated fields, pasture areas and forest areas. Knowledge about the location and limits of cultivated fields, pasture sites and forest sites was obtained through interviews with estate owners and through field excursions. A description of the studied fields and sites is given in Tab. 5.1. To avoid influence from, for example, soil type variations, atmospheric conditions and biomass differences the topographic effect was examined individually for each cultivated field and for each limited pasture and forest site. Data from various cultivated fields, pasture and forest sites were thus not merged.

The dominant crop species were winter wheat, barley, oats,

rape seed, sugar beet, rye and summer wheat. The pasture areas consisted predominantly of moorland with herb vegetation and the uniform woodlands were dominated either by spruce forest or beech forest. The crop cover of the April scene was for some fields very sparse or non-existent, while the cultivated fields of the September acquisition had usually been harvested. In July the fields were dominated by fresh biomass.

5.3.2 Data base

Two types of data were integrated to provide the data base in this study, namely digital LANDSAT data and digital elevation data. LANDSAT MSS (MSS2 and MSS4) as well as LANDSAT TM (TM1-5) data were included. Characteristics of the satellite data used are given in Tab. 5.2. The images were resampled using the cubic convolution algorithm, to fit the national grid of the cadastral map of Sweden (1:10 000). Through this procedure the pixel size was reduced to 50m (MSS pixels) and 25m (TM pixels). A DIPIX ARIES II image analysis system was used for the resampling. Digital elevation data were generated by digitizing the contour lines of cadastral map sheets (5m contour interval). A regular grid net corresponding to the national grid of the cadastral map was thus produced. The digital elevation model (DEM) of each study area was then created by interpolation of the digitized elevation data. An example of a three dimensional plot of one of the DEM used is shown in Fig. 5.2. The interpolation method applied was taken from UNIRAS Software System (European Software Contractors 1982).

The three digital terrain models (two of these with 25m grid size corresponding to TM data and one model with 50m grid corresponding to MSS data) were used for derivation of slope aspect and slope magnitude in degrees for each pixel. Aspect and magnitude were calculated through trigonometry by a moving 3x3 filter over the DEM (DIAPAC software), Olsson and Pinzke 1984). Software was developed to use slope aspect, solar elevation and azimuth to calculate the incidence angle (I) for each pixel. The method used for calculation of the angle of incidence is described by i.a. Holben and Justice (1980) and Smith et al. (1980).

The incidence values for different slope magnitudes on the acquisition occasion of April 27, 1984 are presented in Fig. 5.3. On this date the mean angle incidence value for the studied cultivated fields was 0.721 and varied between 0.823 and 0.530, the mean value for the pasture covers was 0.711 (varying between 0.925 and 0.198) and the mean value for the studied forest covers was 0.740 (varying between 0.949 and 0.098). The solar elevation and azimuth angle configuration of July 6, 1983 make the incidence values at the time for this LANDSAT scene generally higher, while the September image (September 1, 1983, solar elevation 38°, solar azimuth

Table 5.1 Description of the study sites.

	Cultivated field			Pasture cover			Forest cover		
	MSS July	MSS Sept.	TM April	MSS July	MSS Sept.	TM April	MSS July	MSS Sept.	TM April
1	75	75	60	19	19	31	50	50	67
2	4200	4200	13200	1805	1805	20150	4250	4250	34798
3	56	56	220	95	95	650	85	85	522
4	1050	1050	825	451	451	1260	1060	1060	2185

1= number of studied sites, 2= total number of studied pixels, 3= mean number of pixels/site, 4= total studied area (ha).

Table 5.2 Characteristics of the LANDSAT data used.

Acquisition date	Satellite /sensor	Scene	Solar elevation	Solar azimuth
6 July 1983	Landsat 4 MSS	194/21	52°	142°
1 Sept. 1983	Landsat 4 MSS	193/21	38°	150°
27 April 1984	Landsat 5 TM	194/21	44°	148°

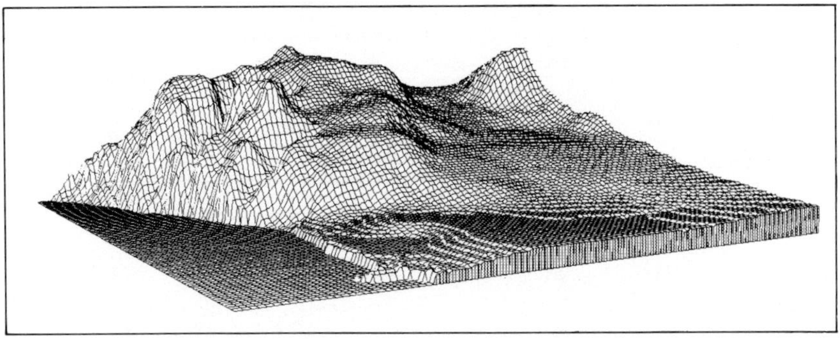

Figure 5.2 Example of a three-dimensional plot of one of the DEM used.

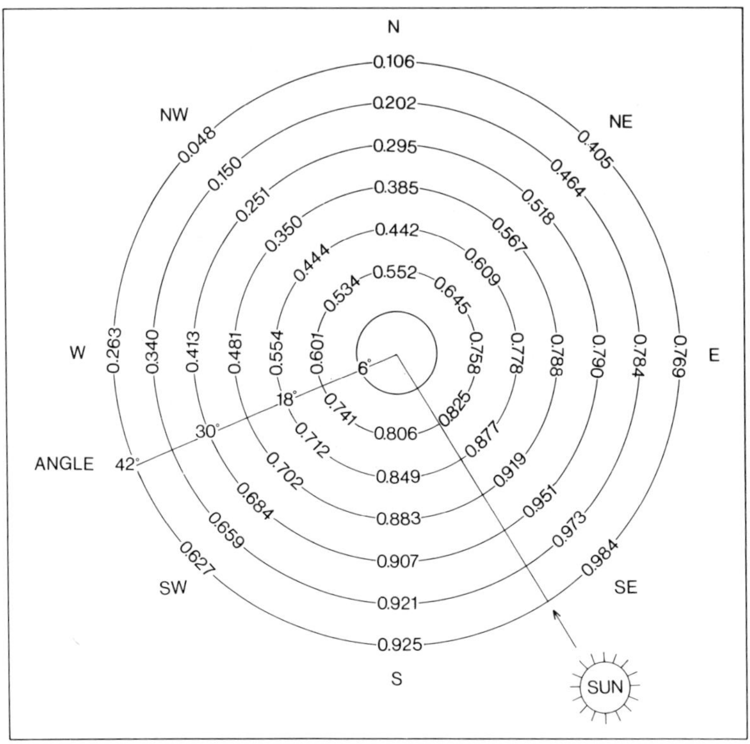

Figure 5.3 A model to show incidence values (cosI) for different aspect strings and slope magnitudes at the acquisirion date of 27 April 1984 (solar elevation 44°, solar azimuth 148.3°). The concentric circles show slope angles, and the numbers are the cosI values, e.g. 0.45.

Table 5.3 The percentage of pixels within each slope magnitude class.

	Cultivated field			Pasture cover			Forest cover		
	MSS July	MSS Sept.	TM April	MSS July	MSS Sept.	TM April	MSS July	MSS Sept.	TM April
1	58	58	56	30	30	24	12	12	15
2	35	35	38	41	41	20	41	41	40
3	7	7	6	24	24	29	38	38	36
4	0	0	0	5	5	7	9	9	9

1= 0.1°-5.0°, 2= 5.1°-10.0°, 3= 10.1°-20.0°, 4>20.1°

Table 5.4 Examples of studied class configurations of incidence angles, slope aspects and slope magnitudes.

Incidence angles (cosI)

A	B	C
0.0 -0.10	0.0 -0.50	0.0 -0.20
0.11-0.20	0.51-1.0	0.21-0.40
0.21-0.30		0.41-0.60
etc.		etc.

Slope aspects (degress)

A	B
337.5- 22.5	292.5- 67.5 (referred to as north facing slopes, N-slopes)
22.5- 67.5	112.5-247.5 (referred to as south facing slopes, S-slopes)
67.5-112.5	
112.5-157.5	
157.5-205.5	
205.5-247.5	
247.5-292.5	
292.5-337.5	

Slope magnitudes (degress)

A	B	C	D	E
0.1-2.0	0.1-10.0	0.1-5.0	0.1-10.0	0.1-5.0
2.1-4.0	>10.1	>5.1	10.1-20.0	5.1-10.0
4.1-6.0			20.1-30.0	10.1-15.0
etc.			etc.	etc.

150.4°) showed lower incidence values. The pixels were dispensed in all aspect strings and there was no dominating slope orientation. The distribution of pixels within different slope magnitude classes is given in Tab. 5.3.

5.3.3. Merging terrain data with LANDSAT data

Analysis of the influence of topography on LANDSAT response was performed by integrating calculated incidence angle values and values of slope aspect and slope magnitude with digital MSS and TM data on a pixel by pixel basis. LANDSAT data as well as terrain data were registered to the national grid net. LANDSAT data were resampled with an average error of less than 0.5 pixels. The integration of satellite data and terrain data was performed by placing data files of angle of incidence, slope aspect, slope magnitude, MSS bands and TM bands as various layers in a data base. The boundaries of each study site marked on the cadastral map were digitized and joined to form another layer in the data base. To eliminate the effect of errors in boundary location and the influence of surrounding pixels, the boundary pixels were removed. The remaining mean number of pixels per site i.e. the mean size of the sites for the three studied cover classes are presented in Tab. 5.1.

Through the DIAPAC software (Olsson and Pinzke 1984) the various layers in the data base were merged and for each study site information about cover class, angle of incidence, slope aspect, slope magnitude and digital grey level (DGL) was obtained on a pixel-by-pixel basis. For each cultivated field, pasture and forest site the relationship between DGL and terrain factors was analysed by linear correlation analysis and analysis of variance. The Pearson product-moment correlation coefficient and the F-test were used (Ryan et al. 1981). Statistically significant differences between data sets refer to the 95% level. Analyses of variance were performed on the classes of incidence angles, slope aspects and slope magnitudes presented in Tab. 5.4.

5.4 RESULTS AND DISCUSSION

Statistical analysis of the relationship between DGL and the angle of incidence, slope aspect and slope gradient revealed the following:

For all acquisition dates and cover classes there was a weak relationship between angle of incidence and DGL. In no cases did the correlation coefficient exceed $\pm$ 0.40.

For all acquisition dates and cover classes there was a low number of sites with statistically significant differences between various incidence angle classes (Fig. 5.4).

For the April and September acquisition there was a low number of cultivated fields with statistically significant differences between N-slopes and S-slopes or between the eight aspect strings (Fig. 5.5.a). (Note that the band width 0.6-0.7μm is named MSS5 and the band width 0.8-1.1μm is named MSS7 in Figs. 5.5.a-c).

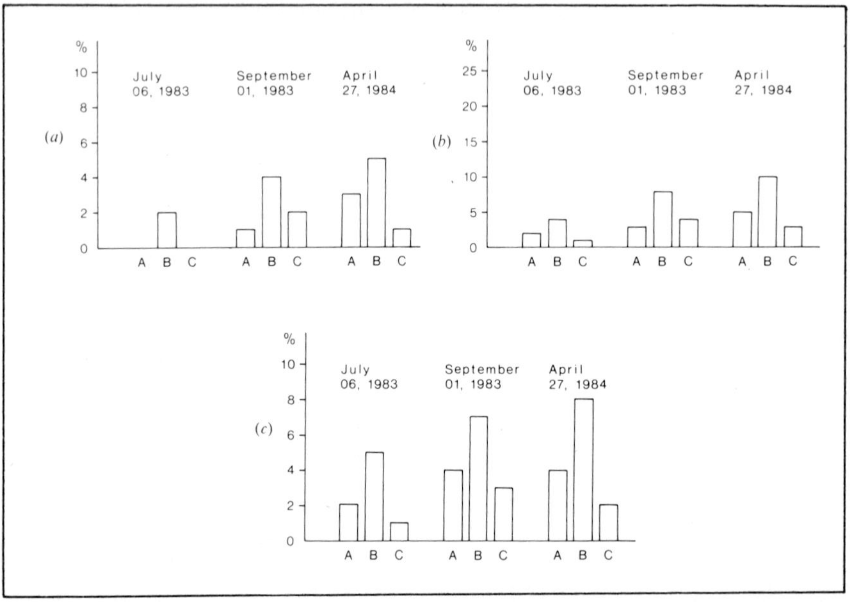

Figure 5.4 The percentage of sites with statistically significant differences (the 95% level) between DGL of various incidence angle classes. A,B,C: various incidence angle configurations (Tab. 5.4). July and September, MSS 4; April TM 5. (a) cultivated fields; (b) pasture sites; (c) forest sites.

For the July acquisition there were no cultivated fields with statistically significant differences between N-slopes and S-slopes or between the eight aspect strings (Fig. 5.5.a).

For the July and September acquisition there was a low number of pasture sites with statistically significant differences between N-slopes and S-slopes or between the eight aspect strings (Fig. 5.5.b).

For the April acquisition 29 per cent (TM4) and 45 per cent (TM5) of the pasture sites showed statistically significant differences between N-slopes and S-slopes (Fig. 5.5.b).

For all acquisition dates there was a low number of forest sites with statistically significant differences between N-slopes and S-slopes or between the eight aspect classes (Fig. 5.5.c).

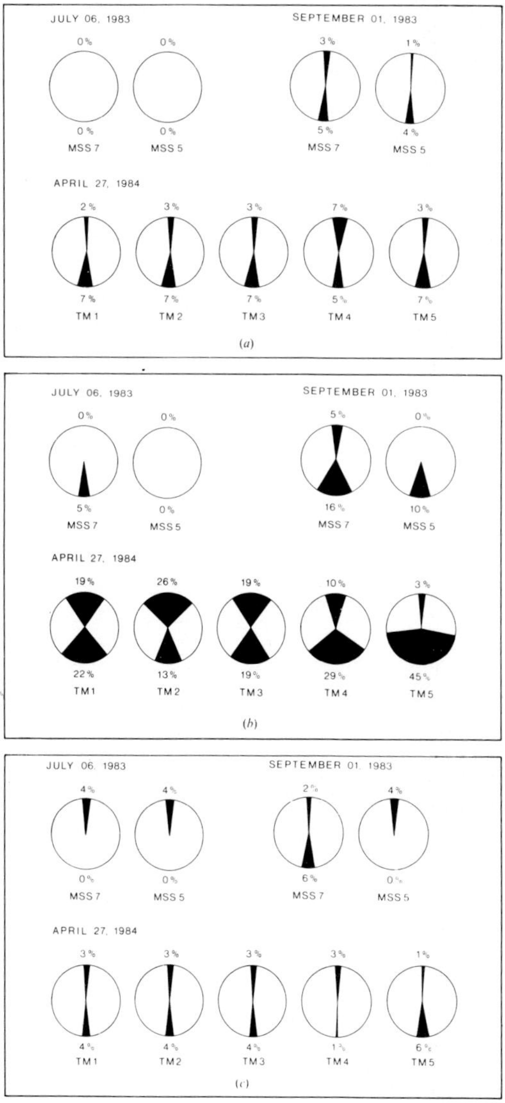

Figure 5.5 The percentage of sites with statistically significant differences (the 95% level) between DGL of N-slopes and DGL of S-slopes. The upper digit, DGL of N-slopes > DGL of S-slopes; the lower digit, DGL of S-slopes > DGL of N-slopes; (a) cultivated fields; (b) pasture sites; (c) forest sites. Note that the band width 0.6-0.7μm is named MSS5 and the band width 0.8-1.1μm is named MSS7.

The test sites with statistically significant differences between N-slopes and S-slopes are concentrated in the higher slope magnitude classes (Tab. 5.5).

Table 5.5 The approximate percentage of sites within each slope magnitude class and the number of sites with statistically significant differences (the 95% level) between DGL of N-slopes and S-slopes within each slope magnitude class.

		6 July 1983				1 Sept. 1983				27 April 1984									
		MSS4		MSS2		MSS4		MSS2		TM1		TM2		TM3		TM4		TM5	
	1	2	3	2	3	2	3	2	3	2	3	2	3	2	3	2	3	2	3
Cultivated fields																			
a	70	0	0	0	0	0	1	0	0	1	0	1	1	1	1	0	2	1	1
b	30	0	0	0		4	1	3	1	3	1	3	1	3	1	3	2	3	1
c	0									0									
Grass and pasture covers																			
a	30	0	0	0	0	0	0	0	0	0	0	0	0	0	0	0	0	0	0
b	40	0	0	0	0	1	1	1	0	2	3	0	1	2	2	2	1	4	1
c	30	1	0	0	0	2	0	1	0	5	3	4	7	4	4	7	2	10	0
Forest covers																			
a	20	0	0	0	0	0	0	0	0	0	0	0	0	0	0	0	0	0	0
b	40	0	1	0	2	1	1	0	2	1	1	1	1	1	1	1	1	2	1
c	40	0	1	0	0	2	0	0	0	2	1	2	1	2	1	0	1	2	0

a= slope magnitude class 0.1-5.0°,
b= slope magnitude class 5.1-10.0°
c= slope magnitude class >10.1°
1= percentage of sites within each slope magnitude class
2= number of sites with DGL of S-slopes > DGL of N-slopes, statistically significant differences, 95%-level
3= number of sites with DGL of N-slopes > DGL of S-slopes, statistically significant differences, 95%-level

For all acquisition dates and cover classes

only negative relationships between DGL and slope magnitude were obtained for N-slopes, while S-slopes demonstrated negative as well as positive relationships;

the strength of the relationship between DGL and slope magnitude did not vary with aspect string;

correlation coefficients between DGL and slope magnitude never exceeded +-0.25 (although for the pasture areas of the April acquisition some correlation coefficients amounted to

+-0.5).

For the April acquisition 15 per cent of the pasture sites indicated statistically significant differences between the slope magnitude classes 0.1°-10.0° and >10.1°. This is valid for TM5 (Fig. 5.6) (with TM1-4 at about 10 per cent). The number of sites with statistically significant differences between various slope magnitude configurations never exceeded 10 per cent for the remaining acquisition dates, cover classes and LANDSAT bands (Fig. 5.6).

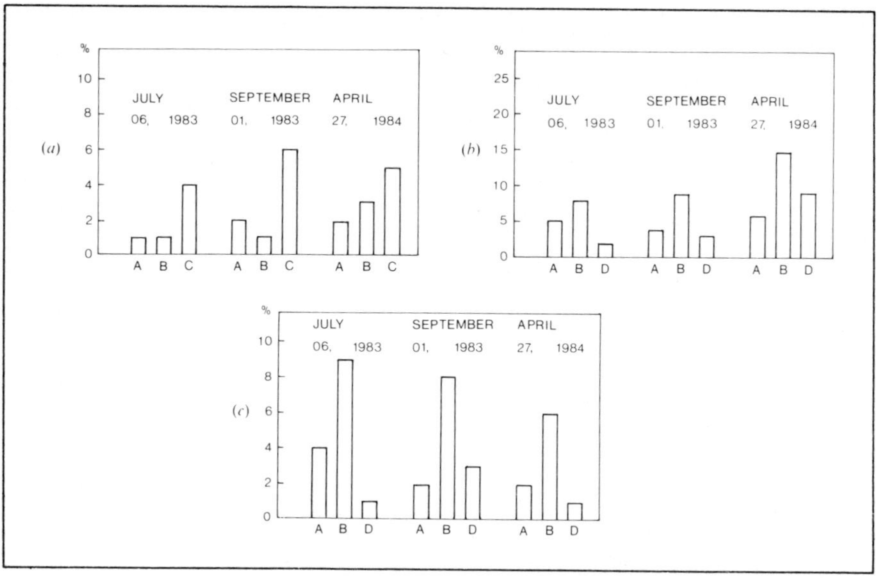

Figure 5.6 The percentage of sites with statistically significant differences (the 95% level) between DGL of vaious slope magnitude classes. A,B,C,D: various slope magnitude configurations (Tab. 5.4). July and September, MSS 4; April TM 5. (a) cultivated fields; (b) pasture sites; (c) forest sites.

The relationship between terrain elements and DGL variations within the investigated regions was thus insignificant for the cover classes cultivated fields and forest areas. This was valid for all studied acquisition dates (solar elevation 52°, 38° and 44°) and for the wavebands MSS2, MSS4 and TM1-5. The LANDSAT response of the pasture areas analysed with July and September data was not found to be related to single terrain parameters, while the April data of these cover types suggested an influence by the topographic conditions.

Some of the forest cover occurs on slopes with very high

slope magnitude (above 50°) and no evident relationship between terrain characteristics and DGL was found for these steep slopes. Several authors have shown that the results of the analysis of covers with these high magnitudes of slope degree are strongly influenced by the topography (Holben and Justice 1980,1981). This unexpectedly low dependence may indicate that the variation of DGL with single terrain elements does not completely describe the possible topographic effect. The weak relationship may thus be explained by the rough geometric structure of the forest cover. This rough structure may cause radiance anisotrophy that "overshadows" the possible relationship between single terrain elements and DGL. The importance of considering the geometric structure of vegetation canopies was elucidated by Smith et al. (1980) who found that the Lambertian assumption, for ponderosa pine, is most valid when the analysis is restricted to slopes of less than 25° and that covers with higher slope magnitudes are better described by non-Lambertian models. It cannot be excluded that the rough forest cover as well as the smoother grass and crop cover would be better described by non-Lambertian models (i.e. models that take into account the geometric structure of vegetation canopies).

5.5 STUDY REMARKS

* Radiometric errors may have arisen due to the resampling.

* The diffuse irradiance that may vary with solar elevation, slope angle, slope aspect and a terrain-view factor that describes the portion of sky obscured by surrounding terrain (Holben and Justice 1980, Estes et al. 1983) was not considered, as were not the possible soil type variations within each study site.

* The amount of biomass per surface area may vary with slope aspect (due to varying amounts of incoming energy and existing soil moisture) and with the magnitude of the slope. This may influence the relationship between DGL and the terrain parameters.

* The geometric structure of the vegetation surfaces was not considered.

* The low number of pixels with high slope magnitudes (above approximately 15°-20°) influences the results of the statistical analysis.

5.6 SUMMARY

Studies of the topographic effect in terrain dominated by slopes between 1° and 15° showed that the relationship between topographic parameters and DGL variations was weak for the cover classes cultivated fields and forest areas.

This was valid for all studied acquisition dates (solar elevation 52°, 38° and 44°) and the wavebands MSS2, MSS4 and TM1-5. The LANDSAT response of the pasture areas analysed with July and September data was not found to be related to terrain parameters, while the DGL variation of the pasture covers of the April imagery was related to slope aspect and slope magnitude.

Cultivated fields in the south of Sweden are most often located in fairly flat terrain. It is concluded that in, for example, qualitative studies of crops, crop forecasting and crop classification within these areas, the effect of topography is of little importance.

The investigation of the relationship between LANDSAT response of natural vegetation surfaces and terrain elements revealed diverging relationships. This cautions us concerning the possibilities of performing detailed investigations of natural cover surfaces within areas of gently undulating terrain before increased knowledge of these relationship is obtained.

Further studies of the topographic effect should concentrate on the interdependencies between the geometric structure of various vegetation surfaces, the satellite response of these surfaces and the topographic conditions.

6

THE RELATIONSHIP BETWEEN LANDSAT DATA AND CROP VARIABLES

Remote sensing as a means of assistance in agriculture includes several fields of application. Two of the most interesting are (i) crop area estimation and (ii) crop yield prediction. Any system to estimate crop production claims the estimate of two factors, namely the yield (kilogrammes per hectare) and the area (hectares). With respect to crop areal estimate, LANDSAT data have proved to be a useful tool.
Crop area estimation demands an accurate and reliable method of classification, which in turn requires comprehension of the relationship between crop type and spectral data. Crop yield prediction needs models based on agrometeorological parameters. Several investigators have indicated the potential of yield prediction models based on agro-meteorological parameters and spectral information of crop conditions.

The purpose of this section was to:

(i) increase the knowledge of the relationship between crop type and spectral data

(ii) study the relationship between at-harvest grain yield, biological yield and spectral data.

Spectral data (vegetation indices) capture the "greenness" or photosynthetic potential of the entire canopies. In this preliminary work spectral data have been used as spectral surrogates of GLAI (Appendix I) in the light-interception subroutines of crop growth. The study will be followed up with detailed studies of the relationship between spectral data and GLAI and with spectral estimates of GLAI as a parameter in agrometeorological growth models.

6.1 BACKGROUND

6.1.1 The interaction of light with vegetation

Section 6.1.1 is based on Myers (1983) and Curran (1985) if not stated otherwise.

The incident energy on the Earth's surface is either reflected, absorbed or transmitted. The proportion between these are different for various types of surfaces and dependent on the characteristics of the Earth's surfaces. It is thus possible to identify features on the Earth's surfaces on the basis of their spectral properties.

For a vegetation surface the proportion between the

reflected, absorbed and transmitted electromagnetic energy varies with wavelength. Fig. 6.1 shows a typical reflectance spectrum of a normal healthy leaf. The basic form of the spectrum is the same for most vegetation types but each is characterized by its own fine structure, the exact position at which peak occurs, and the percentage reflectance in various regions in the spectrum. The spectrum may be divided into three regions:

- The range 0.4μm to 0.72μm: high pigments absorption. The valleys in the visible parts of the curve (Fig. 6.1) are dictated by the pigments in the plant leaves. Higher plants contain four primary pigments, chlorophyll a, chlorophyll b, β caroten and xantophyll. Chlorophyll a and b are the most important and absorb light at wavelengths of 0.43μm and 0.66μm (a) and 0.45μ and 0.65μm (b). Because of the absorption of blue and red energy, healthy vegetation is perceived as green. If the normal growth of the plant for some reason is interrupted, the chlorophyll production may be decreased. The result of this is less chlorophyll absorption in the blue and red and often the red reflectance increases heavily so that the plant turns yellow.

- The range 0.7μm to 1.3μm: high reflectance by morphological structure. The near infrared reflectance is determined by the discontinuities in the refractive indices within a leaf. These discontinuities occur between membranes and cytoplasm within the upper half of the leaf and between individual cells and air spaces of the spongy mesophyll within the lower half of the leaf. The internal structure of plant leaves is mainly influencing the plant reflectance and therefore reflectance measurements enable us to discriminate between species even if they look the same in the visible.

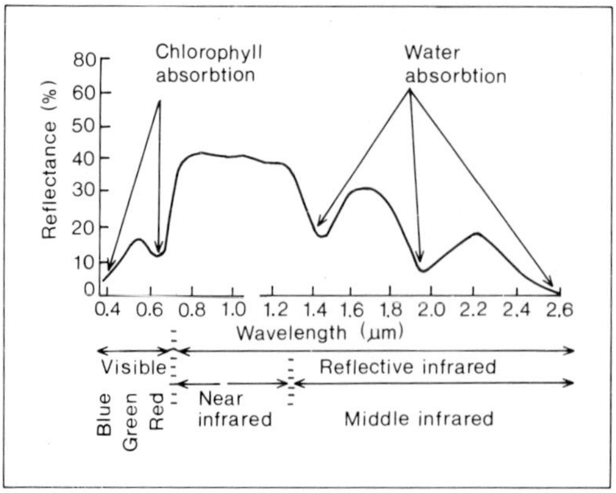

Figure 6.1 Typical reflectance curve of green vegetation (from Lillesand and Kiefer 1979).

- The range 1.3μm to 2.5μm: water absorption. Water absorption is dominating the appearance of the curve in the longer wavelengths.

Since the vegetation canopy is a mosaic of leaves, the reflectance characteristic of a vegetation canopy is more complex than that of a single leaf. Factors influencing the spectral canopy response include:

- amount, arrangement and structure of leaves
- characteristics of other components of the vegetation canopy (stalks, trunks, limbs)
- amount of green, dead or dormant vegetation
- characteristics of the background (soil reflectance)
- solar angle
- look angle
- canopy geometry
- certain episodic and phenological canopy changes
- vegetation productivity
- atmospheric conditions.

Several canopy models have been built which describe canopy reflectance in terms of a number of parameters (Goel and Strebel 1983, Chance and Lemaster 1985).

According to Myers (1983), there is a relationship between the 0.63-0.69μm vegetation canopy radiance and green biomass. This results from the spectral absorption of incident radiation by the chlorophylls. Since the 0.63-0.69 range is inversely proportional to the amount of chlorophyll, it is also sensitive to the amount of photosynthetically active radiation. The relationship between the 0.75-0.80μm near infrared radiance and biomass results from the lack of appreciable spectral absorption and the high degree of intra- and interleaf scattering. The magnitude of difference between the red and near infrared bands increases as the vegetation scene becomes more "green". Conversely, the magnitude becomes less as vegetation becomes dormant or dead (McDaniel and Haas 1982). Several investigators have used this relationship to derive arithmetic functions related to various vegetation variables, e.g. LAI, biomass amount and productivity.

6.1.2 Vegetation indices

Arithmetic functions of different spectral bands are mainly used to (i) perform quantitative measurements of green vegetation, (ii) minimize effects of topography and atmosphere and (iii) separate different classes.

Compared with living green vegetation, dead or dormant vegetation has higher reflectance in the visible ranges and lower in the near-infrared. In the visible, soil has higher reflectance than green vegetation and lower reflectance than dead vegetation. In the near-infrared, soil generally has lower reflectance than green and dead vegetation (Perry and Lautenschlager 1984). Based on these facts, several vegetation indices have been developed to make quantitative estimations of LAI, per cent ground cover, plant height, and biomass.

Two broad categories of vegetation indices have been developed in an attempt to maximize the spectral contribution from green vegetation, while minimizing soil background influences: ratio indices and orthogonal indices. Ratio indices include the near-infrared to red ratio (NIR/red) and the normalized differences (NIR-red/NIR+red). Ortogonal indices include the Kauth and Thomas (1976) greenness vegetation index and the Richardson and Wiegand (1977) perpendicular vegetation index.

Gąrdner et al. (1985) studied the potential value of various TM ratios as vegetation indices;
- very high: TM4/1, TM4/2, TM4/3, TM4/7
- high: TM5/3, TM4/5
- moderate: TM3/1, TM5/1, TM7/1, TM3/2, TM5/2, TM7/2,
- low: TM2/1, TM7/3.

Perry and Lautenschlager (1984) found that several widely used vegetation indices (ratio, normalized difference and transformed normalized difference) are equivalent.

Kauth and Thomas (1976) developed an orthogonal transformation of the original LANDSAT MSS data space to a new four-dimensional space. It was called the "Tasseled Cap" transformation. Each of new axes was intended to be sensitive to different characteristics (e.g. greenness - green vegetation index, brightness - soil brightness index).

Jackson (1983), Crist and Cicone (1984a,b), Crist (1985), Dusek et al. (1985), and Crist and Kauth (1986) have studied and developed the "Tasseled Cap" transformation for MSS- as well as for TM data.

A two-dimensional analogy to the brightness index of the "Tasseled Cap" is the soil line. The perpendicular distance of vegetation to the soil line was used as an indicator of plant development by Richardson and Wiegand (1977).

Several vegetation indices are based on the relationship between infrared reflectance (positive relationship to green biomass) and blue, orange, red reflectance (negative relationship to green biomass). Curran (1982) points out that these relationships are true only until the ground is completely covered by vegetation. After that, further increase in green biomass will result in only small reflectance changes. Tucker et al. (1985) found that the reflectance (NOAA data) is not sensitive to changes below a biomass of 250 kg/ha. Thus, there is a coverage- and biomass range within which reflectance measurements and thereby vegetation indices are most sensitive to biomass variations.

6.1.3 Vegetation identification

Several workers have concluded that spectral measurements at only one time may be unsufficient for the purpose of vegetation discrimination and crop identification. The temporal pattern of spectral data represents the phenological development, i.e. the progress from emergence to maturity of the vegetation canopy. By comparing the pattern of the phenological curve with the pattern of the temporal-spectral curve associated with different vegetation covers, conclusions concerning the vegetation identity may be drawn (Odenweller and Johnson 1984, Ajai et al. 1985).

6.1.4 Before-harvest estimation of at-harvest biomass or yield

Photosynthesis, i.e. the process of conversion of radiant to chemical energy in plants, is the basis of all plant growth. By using sun energy, plants convert carbondioxide and water into carbohydrate which may constitute the harvestable parts of various economically important crops. The rate of the increase of biomass per unit ground area, i.e. the plant or crop productivity, has been related to the amount of solar energy intercepted by the plant (Gallagher and Biscoe 1978). The available light intercepted by the leaves is thus an important factor controlling the productivity of the plants. The amount of light that can be intercepted by the leaves is dependent on the amount of photosynthetically active biomass (GLAI, Appendix I) and its duration (De Wulf and Goossens 1986).

Investigators have found a significant relationship between spectral reflectance measurements and GLAI of crops (Kimes et al. 1981, Ahlrichs and Bauer, 1983, Best and Harlan 1985, Badhwar et al. 1986, Hinzman et al. 1986). However, due to saturation of reflectance very high GLAI cannot be estimated accurately by spectral data.

Curran (1982) suggested that a multispectral reflectance ratio is a measure of vegetation productivity and biomass. The author maintained that a small amount of green biomass with high productivity has a higher reflectance ratio than a high amount of green biomass with low productivity. Tucker and Sellers (1986:1414) concluded that "reflectance data provide indications of the instantaneous biophysical rates associated with plant canopies; gross primary productivity and evaporation, rather than reliable estimates of any state associated with the vegetation, such as LAI or biomass".

When trying to relate reflectance measurements of GLAI to

further rate of plant growth, productivity and final yield, a number of problems arise that might have a significant bearing on the results. One is the assumption that the efficiency with which crops convert radiant energy to biomass is constant. The complex relationships between meteorological factors during various phenological stages and productivity is another example. Uncertainties concerning the relationship between total above-ground biomass and grain yield is a third example. This relationship varies extensively between crop types, regions and climatological conditions. Abiotic and biotic hazards which cannot be predicted or visually assessed in the field may also influence the results.

6.2 REVIEW OF LITERATURE

Some examples of studies concerning the relationship between remote sensing data and:

(i) crop type

(ii) crop or vegetation variables like biomass, yield, green leaf area index and absorption of photosynthetically active radiation are presented.

The number of articles treating these subjects is extensive and therefore the presentation does not claim to be complete. The spectral response of various crop types and the possibility to discriminate various crop types have been studied by for instance Saint et al. (1980), Odenweller and Johnson (1984), Townshend (1984), Ajai et al. (1985), Suits (1985) and others. These studies have mainly been performed with ground-based equipment.

- Using (green x near infrared)/red combinations of ground and airborne measurements, Saint et al. (1980) identified winter crops, spring crops, and annual pasture or forest. For winter crops, a good differentiation was obtained in June and July, between cereals and rape seed. The authors mean that the yellow colour of rape seed flowers has an influence on the channels combination. It was difficult to separate winter wheat and barley in spite of the fact that the sowing time for these crops is not the same. Ratios in April show that the spring growth is comparable while the decrease of activity for barley is faster than for wheat which is still active (in June). In July, barley maturity is earlier than wheat maturity. For spring crops it was possible to identify corn, sorghum, and soy-bean, with two observation dates. Activity of corn and soy-bean is higher than sorghum in June. In September, corn and sorghum activity is decreasing while soy-bean, which is a late crop still has high activity.

- Odenweller and Johnson (1984) found that temporal-spectral profiles (Appendix I) support crop identification. The authors could separate individual crops (corn and soy-beans) on the basis of amplitude and distinctive shape of the profiles.

- Townshend (1984) discriminated agricultural land-cover (sugar-beet, orchards, grassland, senescent peas, straw and stubble, bare soil) using thematic mapper spectral bands. Results indicated that an optimal choice of bands should include bands 3, 4, and a band from the middle- or far-IR.

- The possibility to discriminate three winter crops, namely wheat, chickpeas and mustard, on the basis of their temporal-spectral profiles was examined by Ajai et al. (1985). Using various profile features the three crops growing in the same season could be identified.

- A row crop reflectance model was developed by Suits (1985) in order to compare the soy-bean - corn discrimination potential of MSS bands and TM bands in mid-July and in mid-August. The reflectance model confirms that TM bands can be used earlier in the season than the MSS bands and implies that the sensitivity of the TM mid-IR bands to exposed soil between rows of the early soy-bean canopy may cause the early discrimination potential.

The relationship between vegetation biomass or crop yield and spectral data has been studied by Pearson et al. (1976), Bauer et al. (1981), Barnett and Thompson (1982), Aase et al. (1984), Patel et al. (1985), Ripple (1985a,b), Peden et al. (1986).

- Pearson et al. (1976) performed measurements of gramineous biomass using a hand-held instrument. A good relationship between biomass and spectral data was obtained but the authors underline that the vegetation canopy should contain at least 30% live or green vegetation.

- Basing their study on several thousand reflectance spectra of maize, soy-bean and wheat canopies, Bauer et al. (1981) concluded that yields <u>cannot</u> directly be predicted from spectral measurements.

- Radiance measurements by a hand-held radiometer and an airborne multispectral scanner were related to wheat stand densities and to grain yield by Aase et al. (1984). Correlation analyses between the IR/R ratio and grain yield (t/ha) produced coefficients between 0.8 and 0.9.

- Final grain yield and straw yield on various varieties of rice were related to ground-based radiometric measurements by Patel et al. (1985). The correlation coefficient between yield and spectral parameters was highest (r=0.78) around the heading stage of the crop.

- Ripple (1985a) studied the relation between LANDSAT TM bands (Barnes Modular Multiband Radiometer) and various grass canopy characteristics. TM band 4 yielded r=0.80 between total wet biomass and spectral data. The same author (Ripple 1985b) showed that the relationships between spectral bands and canopy variables were curvlinear for the biomass range reported in the article (total wet biomass range 16.5-1677.9 g/m^2 and total dry biomass range 6.5-502.5 g/m^2). Spectral bands showed a sensitivity to total wet and dry biomass as well as to plant water and LAI. TM5 was the least sensitive band compared with TM1-4 and TM6-7.

- Highly significant relationships (R-sq.= 0.77) between yield of maize (t/km^2) and the NIR/R ratio (aerial photography) were obtained in an investigation carried out by Peden et al. (1986).

The possibility to estimate green leaf area index (GLAI) by spectral measurements have been studied by Wardely and Curran (1984), Best and Harlan (1985), Hatfield (1985), and Curran and Williamson (1987a,b).

Steven et al. (1983), Daughtry et al. (1984), Gallo et al. (1985) measured the absorption of photosynthetically active radiation (PAR).

- Steven et al. (1983) used a two-band spectrometer to measure solar radiation reflected from sugar beets. Measurements of light interception was performed by placing solarimeters at the base and above the crop. Leaf cover was determined from photographs over the crop. The authors found the light interception to be proportional to leaf cover and that the relationship was almost 1:1. The ratio IR/R increased with the percentage of ground covered by healthy green leaves and was used to determine the fraction of incident solar energy intercepted by the leaves. Measurements were used to predict yield to within 6 per cent.

- The fraction of absorbed PAR by plants was estimated from measurements of visible and near-infrared canopy reflectance by Asrar et al. (1985b). In conjunction with micrometeorological data these measurements were used to estimate above-ground total dry phytomass (g/m^2). Strong correlations were obtained between estimated and measured phytomass.

- Gallo et al. (1985) developed relationships between PAR absorbed by corn and the spectral response of the canopies. The photosynthetic photon flux densities entering and exiting the corn canopies were measured with a line quantum sensor. Absorbed PAR, cumulated through the season was found to produce a closer relationship with yield than the cumulated LAI.

6.3 APPROACH

6.3.1 Material

Remote sensing data included:

- LANDSAT MSS acquisitions of the province of Scania (1983; May 19, June 20 and July 06)

- LANDSAT TM acquisitions of the province of Östergötland (1984; June 07, July 09 and August 26).

The relationship between satellite data and crop factors was studied and the results are reported in this section. Concerning the province of Scania, crop factors included (i) crop type and (ii) crop yield (kg/ha) while the crop factors associated with the province of Östergötland were (i) crop type and (ii) crop biological yield (g/m^2).

- The province of Scania (MSS acquisitions).

Information concerning the location of fields and the crop type and yield of these fields was obtained through interviews with farmers. The estates were selected by simply picking names from a telephone directory. Through a letter hundreds of individual estate owners, in all parts of Scania, received information about the project and were asked to contribute. Farmers who did not want to participate were against the idea of giving information to a research student without an established organisation. Some farmers referred to the possibility of control and did therefore not contribute while others considered their data not very accurate. Estate owners that accepted were visited. During these visits, the location of each field of the farm was marked on a map and the crop type, total end of season yield and field area was noted. For some fields, information concerning e.g. fertilization and diseases was obtained. Usually the estate owners had very precise historical records of the location of various crops as well as yield. The accuracy of the crop data may therefore be considered to be good. In spite of the fact that the data were not strictly randomly sampled they were handled as a statistical sample. A presentation of crop types, number of fields and field sizes is given in Tab. 6.1.a.

- The province of Östergötland (TM acquisitions).

Crop data were received through cooperation by the Central Bureau of Statistics (SCB) and consisted of information concerning (i) crop type and (ii) biological yield (i.e. for cereals - the total above ground dry substance, g/m² and for potatoes the underground potato mass, kg/ha).

SCB performs each year an objective yield prediction. In this a number of fields are randomly selected. For cereals the biological yield of two 1m² plots within each selected field is measured. This is performed less than three weeks before harvest. For potatoes the biological yield of exactly 2m of one row is measured (two areas within each field). These measurements are transformed to kg/ha.

Table 6.1.a Crop data for the province of Scania.

	No. of fields	Field size (ha)			Grain yield*			
		mean	max	min	mean	s.	min	max
Winter wheat	46	35	103	5	56	9	37	80
Summer wheat	32	19	50	4	38	7	20	50
Rye	33	21	60	7	37	4	32	47
Barley	43	25	73	5	35	7	20	48
Oats	35	20	57	6	35	9	17	52
Peas	31	26	66	8	36	8	22	44
Grazing crop	41	34	64	16	76	11	67	86
Rape seed	36	30	55	4	32	7	16	43
Sugar beet	20	29	79	5	353	56	250	416

Table 6.1.b Crop data for the province of Östergötland.

	No. of plots (1m²)							Biological yield**				
area:	1	2	3	4	5	6		T.	mean	s.	min	max
Winter wheat	29	110	41	63	60	31	=	334	494	179	31	960
Summer wheat	-	26	15	-	9	10	=	60	460	119	188	736
Rye	23	61	38	25	53	12	=	212	421	102	98	676
Barley	-	108	35	54	63	34	=	294	437	140	17	931
Oats	-	70	20	19	33	27	=	169	496	119	109	705
Potatoes***	-	46	62	-	53	-	=	161	478	97	232	670
Grazing crop	-	48	22	16	33	25	=	144	467	158	68	824

*= (10²kg/ha), water content - cereals 15%, grazing crop 16.5%, rape seed 18%.
**= (g/m²), dry matter.
***= (kg/ha).
T= total
s= standard deviation

The crop data were obtained from SCB in coordinated form i.e. each plot ($1m^2$) with information about crop type and biomass was described by x-and y-coordinates (the national grid of Sweden). Thus, the exact location of each plot was known. Tab. 6.1.b gives information about these crop data.

6.3.2 Methods

- MSS data

The fields of Scania (Tab. 6.1.a) were identified by means of field maps at an EBBA colour video monitor and the preprocessed MSS data (not resampled) of each field were measured at the EBBA system.

- TM data

Preprocessed TM data and crop data were integrated by developing software that from the x-and y-coordinates in the crop data file read the TM pixel (resampled 20m pixel) overlaying the $1m^2$ plot. Thus, the resultant file comprised for each x- and y-coordinate information concerning crop type, biological yield and satellite data.

6.3.3 Purpose and procedure

The objective of this section was to study:

1. The relationship between crop type and satellite data (MSS and TM data).

2. The relationship between crop yield and satellite data (MSS data).

3. The relationship between data of the objective yield predictions, performed by SCB (in the form of biological biomass) and satellite data (TM data).

The aim of point 3 was not to study the possible influence of biological yield on satellite data. The ground data obtained from SCB (two $1m^2$ plots within each field) are not suitable for this purpose. Curran and Williamson (1985, 1986) discuss the accuracy of ground data in remote sensing studies. The purpose of point 3 is associated with the idea of replacing a number of test fields in the objective yield predictions by satellite measurement. The aim of point 3 is therefore rather to get a very preliminary estimation of the possibility and value of performing replacements like these.

Point 1 was performed by:

- Studying the statistical significance between various crop type groups and individual crops on a monotemporal basis (analysis of variance, the F-test).

- Examining the multitemporal curve of each crop type.

Point 2 included:

- Linear analysis of the relationship between MSS data and crop yield (correlation coefficient analysis).

- Non-linear analysis of the relationship between MSS data and crop yield (correlation coefficient analysis);
  - log spectral data
  - log crop data
  - log spectral data and log crop data.

- Analysis of variance between biological yield classes (F-test).

Point 3 included:

- Linear as well as non-linear analysis of the relationship between TM data and objective yield data (correlation coefficient analysis).

- Analysis of variance between biological yield classes (F-test).

TM data comprised six sub-areas (section 4) and for each field two $1m^2$ plots were available. Analyses of TM data (point 1 and 3) were therefore performed according to the following options (i) for each $1m^2$ plot in each sub-area, (ii) for each $1m^2$ plot in all sub-areas together, (iii) for the mean of two $1m^2$ plots (belonging to the same field) in each sub-area and (iv) for the mean of two $1m^2$ plots in all sub-areas together.

Each of the points (1-3) and their sub-matters was performed on:

a. Not atmospherically corrected data

b. Haze removed data (section 4)

c. Standardized data (section 4)

d. Ratios

a. Not atmospherically corrected data

Not atmospherically corrected data included:

- spectral radiance ($L\lambda$)
- at-satellite spectral reflectance ($\rho P\lambda$)

b. Haze removed data

Haze removal was performed on DN-values and then the data were transformed to:

- spectral radiance (Lλ)
- at-satellite spectral reflectance ($\rho P\lambda$)

which were used in points 1-3

c. <u>Standardized data</u> (section 4)

- At-satellite spectral reflectance ($\rho_{p\lambda}$) and
- Bλ,s (sections 3 and 4)

were used in the standardization approach.

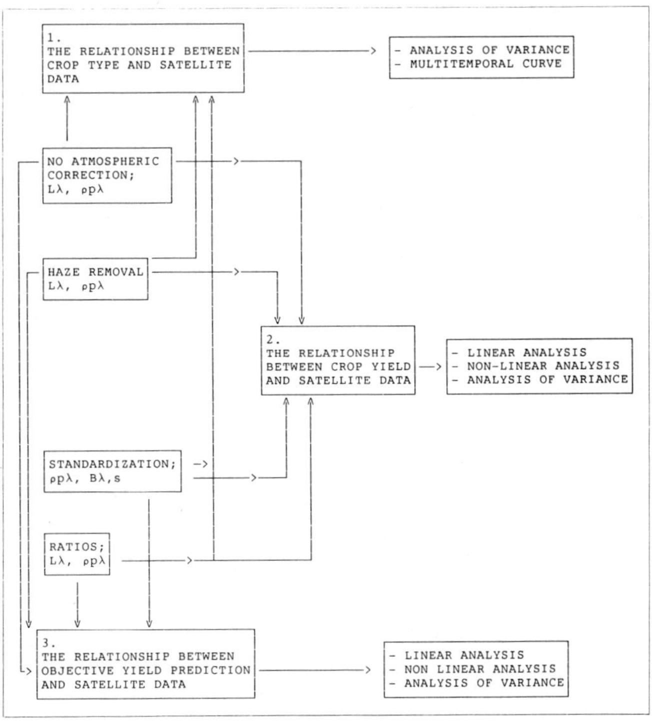

<u>Figure 6.2</u> Summation of the applied procedures.

d. Ratios

Simple division of bands were performed on:

- spectral radiance ($L\lambda$)
- at-satellite spectral reflectance ($\rho P\lambda$)

and based on points a,b and c above

The following ratios were used in points 1-3:

- MSS 4/2
- TM 4/3, 7/3, 4/7, 4/5, 5/3 and ND (= 4-3/4+3).

For points (1-3) the statistical significance of results obtained by the four approaches (a-d, above) was studied. The applied procedures are summarized in Fig. 6.2.

6.3.4 Statistical analysis

Statistical analysis included calculation of the Pearson product-moment correlation coefficient, t-test of statistically significant correlation coefficients and the F-test (analysis of variance) (Ryan et al. 1985). Statistically significant differences refer to the 95% level. The requirement that the input data should be normally distributed was tested. Integrations of crop data and satellite data and analyses of relationships were performed on a microvax system.

6.4 RESULTS

6.4.1 Crop type

- Monotemporal studies.

Some examples of results of analyses of variance between crop types are given in Tabs. 6.2-6.5 and in Appendix A.

Tab. 6.2 - crop type against crop type, MSS data, not atmospherically corrected ($\rho P\lambda$).

Tab. 6.3 - winter against summer crop, TM data, standardized ($\rho P\lambda$), the mean of two $1m^2$ plots (of the same field) for individual sub-areas.

Tab. 6.4 - winter against summer crop, TM data, standardized ($\rho P\lambda$), the mean of two $1m^2$ plots (of the same field) and all sub-areas together.

Appendix A - crop type against crop type, TM data, standardized ($\rho P\lambda$), each $1m^2$ plot and each sub-area.

Table 6.2 Statistically significant differences between individual crop types.

	S. WHEAT			RYE			BARLEY			OATS			PEAS			GR.CR.			RAPE SEED			SUGAR BEET		
MSS:	4	2	4/2	4	2	4/2	4	2	4/2	4	2	4/2	4	2	4/2	4	2	4/2	4	2	4/2	4	2	4/2
1. W.WHEAT																								
MAY 19	Y	Y	Y	Y	N	N	Y	Y	Y	Y	Y	Y	Y	Y	Y	Y	N	N	Y	Y	Y	Y	Y	Y
JUNE 20	Y	Y	Y	Y	Y	Y	Y	Y	Y	Y	Y	Y	Y	Y	Y	Y	Y	Y	Y	Y	Y	Y	Y	Y
JULY 06	Y	Y	Y	Y	Y	Y	Y	Y	N	Y	N	N	N	Y	N	N	N	Y	Y	Y	Y	Y	Y	Y
2. S.WHEAT																								
MAY				N	Y	Y	N	N	N	N	N	Y	N	N	Y	Y	Y	Y	Y	Y	Y	N	N	Y
JUNE				Y	N	Y	N	N	Y	N	N	Y	N	N	Y	N	Y	Y	Y	Y	Y	Y	Y	Y
JULY				Y	Y	Y	N	N	N	N	N	N	N	Y	N	Y	N	N	Y	Y	Y	Y	Y	Y
3. RYE																								
MAY							N	N	N	N	N	N	Y	Y	Y	Y	Y	Y	Y	Y	Y	N	N	N
JUNE							Y	N	Y	Y	N	N	Y	Y	N	Y	Y	Y	Y	Y	N	Y	Y	Y
JULY							Y	Y	Y	Y	Y	Y	Y	N	Y	Y	N	Y	Y	N	Y	Y	Y	Y
4. BARLEY																								
MAY										N	N	N	N	N	N	Y	Y	Y	Y	Y	Y	N	Y	Y
JUNE										Y	N	Y	N	Y	Y	N	Y	Y	Y	Y	Y	Y	Y	Y
JULY										N	N	N	N	Y	N	Y	N	N	Y	Y	Y	Y	Y	Y
5. OATS																								
MAY													N	N	N	Y	Y	Y	Y	Y	Y	N	Y	Y
JUNE													N	N	N	N	N	N	Y	N	Y	Y	Y	Y
JULY													N	N	N	Y	N	Y	Y	Y	Y	N	Y	Y
6. PEAS																								
MAY																Y	Y	Y	Y	Y	Y	N	N	N
JUNE																N	N	N	Y	N	Y	Y	Y	Y
JULY																N	N	N	N	N	N	N	Y	Y
7. GRAZING CROP																								
MAY																			N	N	N	Y	Y	Y
JUNE																			Y	Y	Y	Y	Y	Y
JULY																			Y	N	Y	Y	Y	Y
8. RAPE SEED																								
MAY																						Y	Y	Y
JUNE																						Y	Y	Y
JULY																						Y	Y	Y

Y= statistically significant difference (the 95% level)
N= not statistically significant difference (the 95% level)

Table 6.3 Statistically significant differences between winter and summer crops.

TM:	3	4	5	7	4/3	4/7	5/3	4/5	7/3	3	4	5	7	4/3	4/7	5/3	4/5	7/3	3	4	5	7	4/3	4/7	5/3	4/5	7/3
	AREA 1									AREA 2									AREA 3								
JUNE 07	N	Y	Y	Y	Y	Y	Y	Y	N	Y	Y	Y	Y	Y	Y	Y	Y	N	N	Y	Y	N	Y	Y	Y	Y	N
JULY 09	Y	Y	Y	Y	Y	Y	Y	Y	N	Y	Y	Y	Y	Y	Y	Y	Y	N	Y	Y	Y	Y	Y	Y	Y	Y	Y
	AREA 4									AREA 5									AREA 6								
JUNE 07	N	Y	N	Y	Y	Y	Y	Y	N	Y	Y	Y	Y	Y	Y	Y	Y	N	N	Y	Y	N	N	Y	Y	Y	N
JULY 09	N	Y	N	Y	Y	Y	Y	Y	N	Y	Y	Y	Y	Y	Y	Y	Y	N	Y	Y	Y	Y	Y	Y	Y	Y	Y

Y= statistically significant difference (the 95% level)
N= not statistically significicnt difference (the 95% level)

Table 6.4 Statistically significant differences between winter and summer crops.

TM:	3	4	5	7	$\frac{4}{3}$	$\frac{4}{7}$	$\frac{5}{3}$	$\frac{4}{5}$	$\frac{7}{3}$
JUNE 07	N	Y	Y	Y	Y	Y	Y	Y	N
JULY 09	Y	Y	Y	Y	Y	Y	Y	N	

Y= statistically significant difference (the 95% level)
N= not statistically significant difference (the 95% level)

Table 6.5 Summation of Appendix A. The crop types producing statistically significant differences in all sub-areas (1-6) of Östergötland.

June 07	July 09
winter wheat - potatoes	winter wheat - potatoes
winter wheat - rape seed	winter wheat - rape seed
summer wheat - rye	summer wheat - rye
summer wheat - rape seed	summer wheat - potatoes
rye - barley	summer wheat - rape seed
rye - potatoes	rye - barley
rye - rape seed	rye - oats
barley - potaoes	rye - potatoes
barley - rape seed	rye - grazing crop
oats - potatoes	rye - rape seed
oats - rape seed	barley - potatoes
potatoes - grazing crop	barley - grazing crop
grazing crop - rape seed	barley - rape seed
	oats - potatoes
	oats - grazing crop
	oats - rape seed
	potatoes - grazing crop
	grazing crop - rape seed

Tab. 6.5 - Summation of Appendix A.

Statistically significant differences were obtained between the winter and summer crops of cereals for all acquisition dates.

Individual cereals like winter wheat, summer wheat, rye, and oats seldom expressed statistically significant differences.

Individual crops yielding statistically significant differences with the other crop types on all acquisition dates were (i) potatoes, (ii) rape seed, (iii) sugar beets.

In July rye differed statistically from summer wheat, barley, and oats.

Results of the analyses, based on the four different approaches: (i) no atmospheric correction, (ii) haze removal, (iii) standardization and (iv) ratio, point a-d, section 6.3.3 as well as results based on various approaches like spectral radiance, at-satellite spectral reflectance etc. did not express any comprehensive variation considering the presence of statistically or not statistically significant differences.

- Multitemporal studies.

Multitemporal profiles based on the mean value of each crop type were created. For each crop type analysis of variance was performed between dates. These differences were often significant and error bars have not been included in the diagrams presented below.

Some examples of multitemporal profiles are presented in Figs. 6.3-6.5;

Fig. 6.3.a winter- summer crop, TM4 (L),
Fig. 6.3.b winter- summer crop, TM4 (ρP), standardized
Fig. 6.3.c winter- summer crop, TM5 (L)
Fig. 6.3.d winter- summer crop, TM5 (ρP), standardized
Fig. 6.4.a individual crops, MSS4 (ρP), haze removed
Fig. 6.4.b individual crops, MSS2 (ρP), haze removed
Fig. 6.5.a individual crops, TM4 (L)
Fig. 6.5.b individual crops, TM4 (ρP), standardized
Fig. 6.5.c individual crops, TM5 (L)
Fig. 6.5.d individual crops, TM5 (ρP), standardized
(Figs. 6.3.a-d, 6.5.a-d - TM data are valid for each 1m^2 plot, all sub-areas together).

In Fig. 6.3.a (TM4, L) all winter crops together express a downward trend from the beginning of June to the end of August while the summer crops show a rising between June and July and then a decline to August.

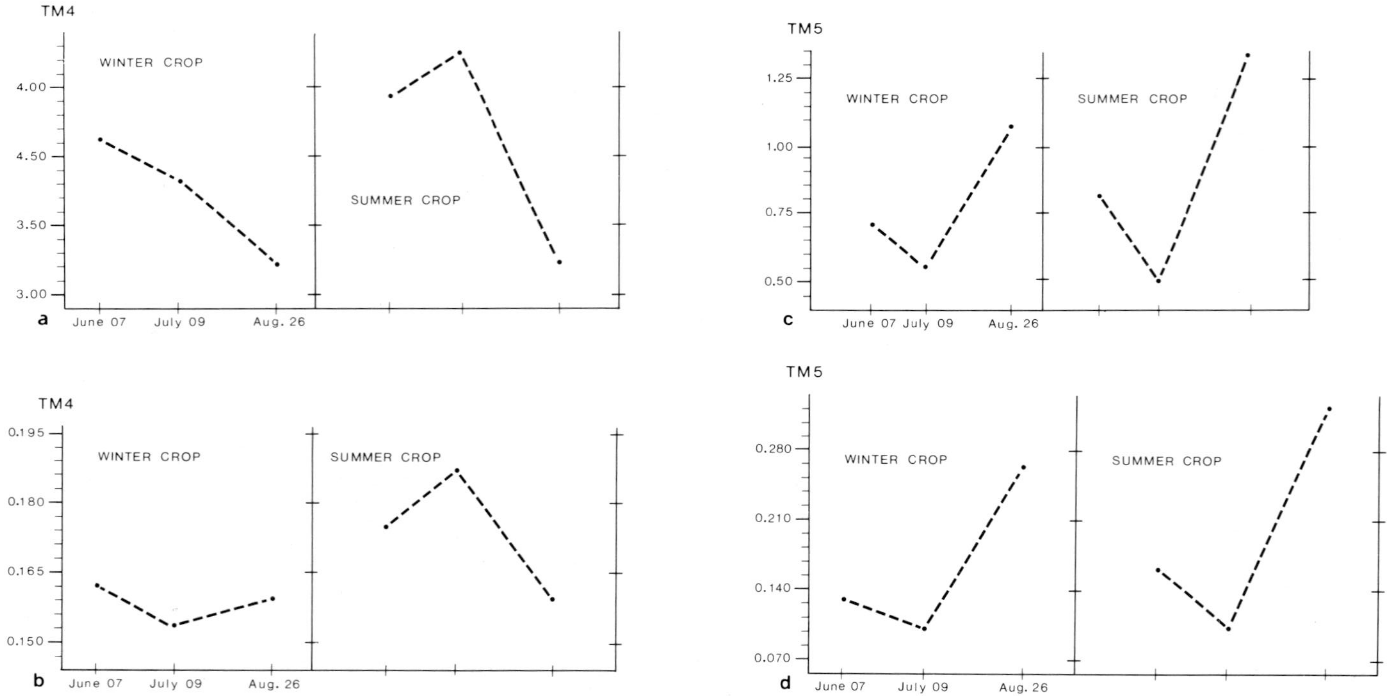

Figure 6.3. TM data of winter and summer crops plotted against acquisition date.
(a) TM4, the mean spectral radiance
(b) TM4, the mean at-satellite spectral reflectance, standardized
(c) TM5, the mean spectral radiance
(d) TM5, the mean at-satellite spectral reflectance, standardized

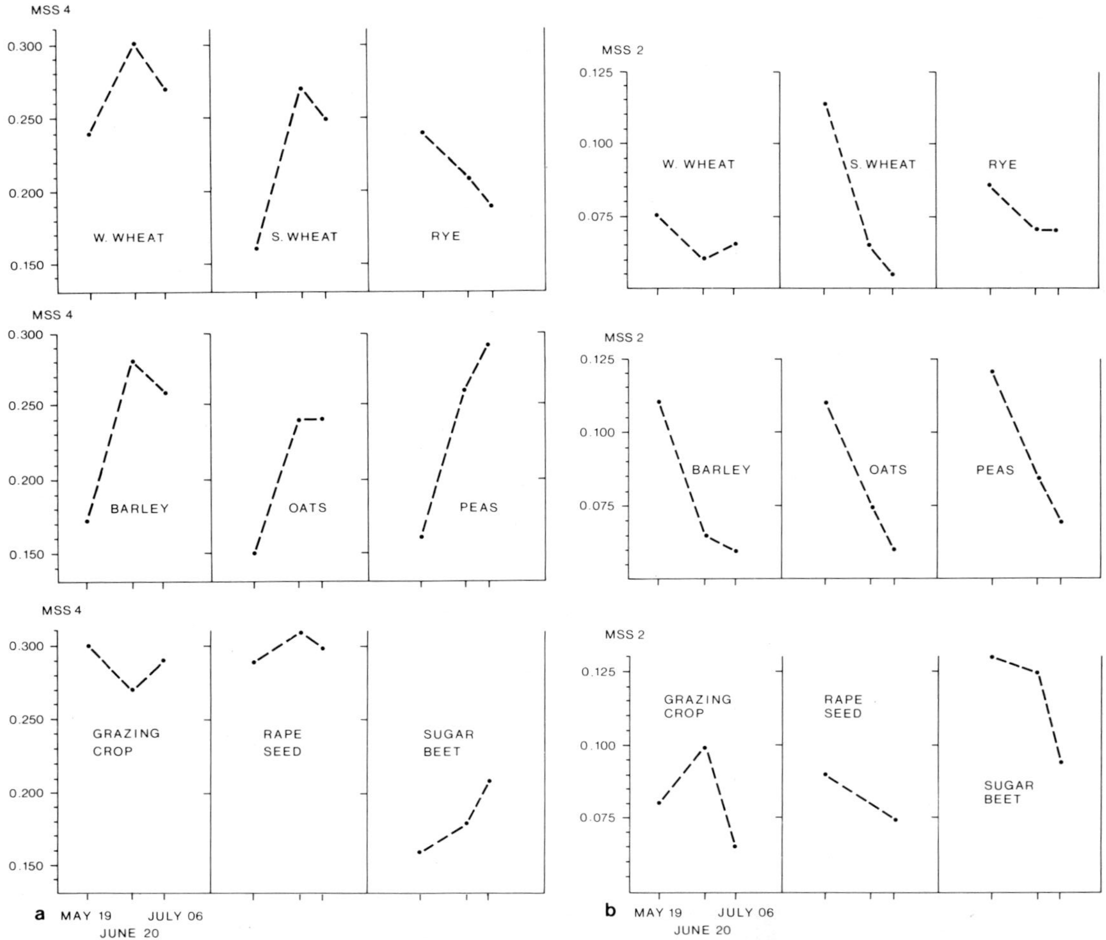

Figure 6.4
The mean at-satellite spectral reflectance of individual crops plotted against acquisition date.
(a) MSS4
(b) MSS2

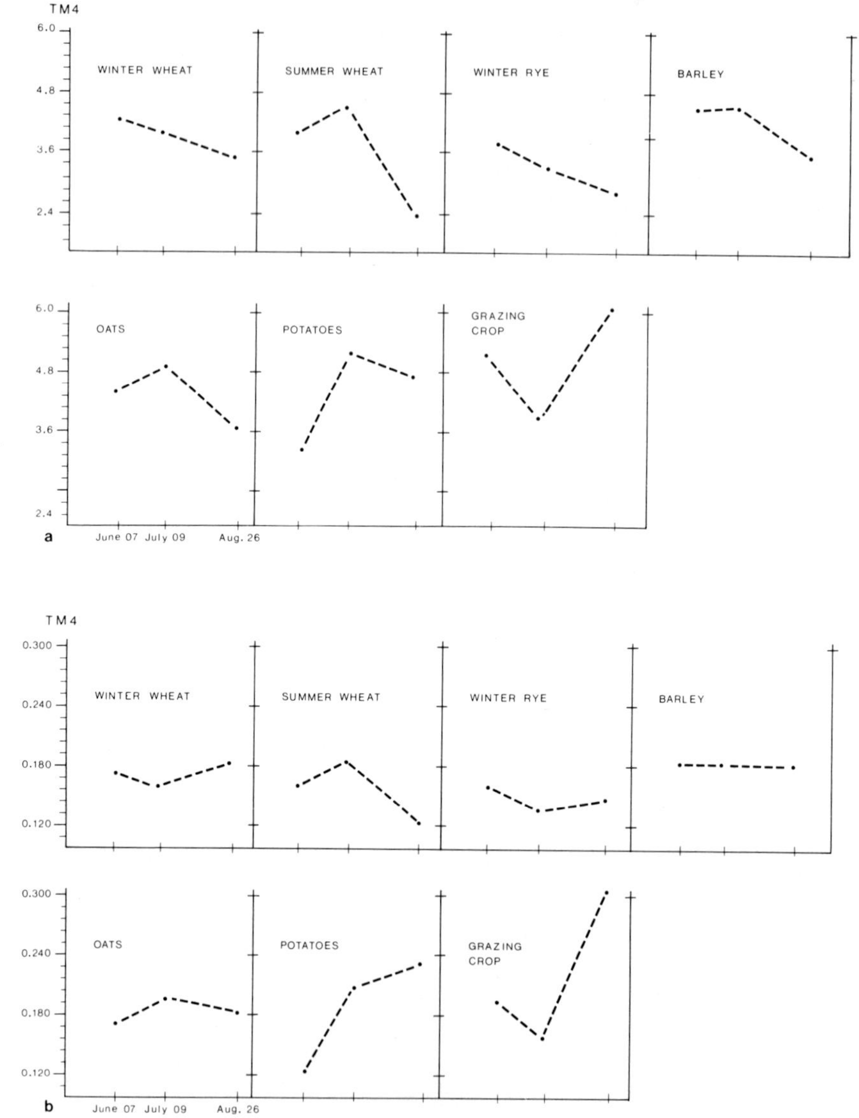

Figure 6.5 Spectral data of individual crops plotted against acquisition date.
a= TM4, spectral radiance,
b= TM4, at-satellite spectral reflectance

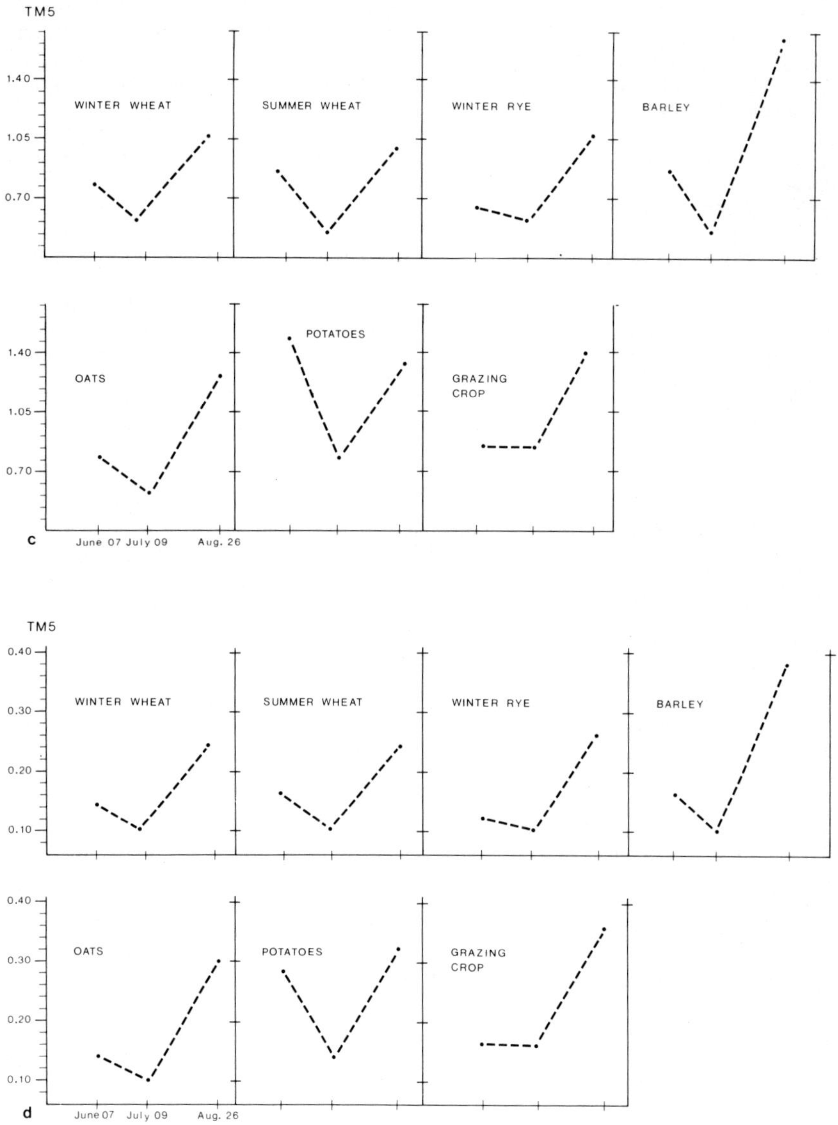

Figure 6.5 cont.
c= TM5, spectral radiance,
d= TM5, at-satelitte spectral reflectance.

When comparing Fig. 6.3.a with Fig. 6.3.b (TM4, ρP), similar results were obtained for summer crops but not for winter crops, which produced a decrease between June and July and a rise between July and August.

Fig. 6.3.c (TM5, L) and Fig. 6.3.d (TM5, ρP) yielded similar profiles, i.e. a diminution between June and July and a rise between July and August.

Winter wheat, summer wheat, barley, oats, peas, rape seed and sugar beets produced increased spectral radiance values of haze removed MSS4 data between the 19th of May and the 20th of June, while rye and grazing crops expressed a decrease. Between the 20th of June and the 6th of July peas, grazing crops and sugar beets increased their ρP-values of haze removed data while the other crops reduced theirs (Fig. 6.4.a). MSS2 (Fig. 6.4.b) normally showed opposite trends to MSS4.

Figs. 6.5.a-d present the individual profile (TM data) of each crop type. When comparing for example Fig. 6.5.a (spectral radiance, TM4) with Fig. 6.5.b (at-satellite spectral reflectance, TM4) it is seen that the appearance of the profiles was influenced by which preprocessing alternative that was used ($L\lambda$, $\rho P\lambda$, $B\lambda,s$, not atmospherically corrected, haze removed, standardized etc.).

When comparing Fig. 6.5.b (ρP, standardized) with ρP data not standardized, comparable trends were obtained. It was therefore concluded that the differences between Figs. 6.5.a and 6.5.b are due to a difference between spectral radiance and at-satellite reflectance rather than to the atmospheric influence minimizing procedure. The increase of for example ρP,TM4 (Fig. 6.5.b) between July 09 and August 26 for winter wheat indicates an increase in "freshness" of the crop. This is certainly rather remarkable and the conclusion drawn from this may either be that the at-satellite reflectance is unsuitable as a sun elevation correction approach (e.g. overcompensates for the low sun angle in August) or that the fields are already grown with some new green vegetation (quite unlikely). In fact, it was frequently found that the sun elevation correction methods applied caused deviations from that what was "expected".

The mono- and multitemporal crop type studies showed that:

- Through the vegetation period winter crops differed significantly from summer crops.

- Rape seed, potatoes and sugar beets differed significantly from the other crops. According to Saint et al. (1980), the yellow colour of the flowers has an influence on the remote sensing data. Potatoes differ from cereals in growing geometry, while sugar beets have a unique phenological curve.

- Rye often has a bluish-green colour and it turns yellow earlier than the other crops. This probably influences the clear separability that was obtained between rye and other cereals.

- Barley (summer crop) is a short season crop and ripens quickly, which was seen in some multitemporal profiles. This crop was often confused with winter wheat. This was also found by Saint et al. 1980.

- For TM data similar results were obtained when calculating with each 1m² plot as with the mean of two 1m² plots (in the same field). Results of various sub-areas varied (Appendix A). The fundamental features were however the same as they were in all sub-areas together.

- The appearence of the temporal-spectral profiles and the conclusions based on these may vary with solar elevation correction approach and whether a solar correction was applied or not.

- The applied atmospheric minimizing procedures were not found to influence significantly the mono- and multitemporal studies. The conclusions of the results based on the various approaches are therefore similar. This is also valid for ratios compared with single bands.

Results indicate a promising possibility to separate crop type groups and certain individual crop types. It is therefore considered worthwile to perform studies of crop classification potentials.

6.4.2 Crop yield

Appendix B gives the Pearson product-moment correlation coefficients between MSS data acquired at three dates and at-harvest grain yield (kg/ha). A reference to a scatter plot of the relationship is also presented. These tables are valid for haze corrected spectral radiance ($L\lambda$). Various preprocessing alternatives did not produce correlation coefficients with statistically significant differences.

According to Rudberg (1975), a correlation coefficient between $\pm$ 0.40 and $\pm$ 0.69 indicates a <u>fairly</u> strong relationship, while a correlation coefficient between $\pm$ 0.70 and $\pm$ 0.99 indicates a <u>very</u> strong relationship. Correlation coefficients higher than 0.70 were never obtained but several cases with coefficients higher than 0.40 were present (Appendix B). A confidence interval test (the 95% level) revealed that the cases presented in Tab. 6.6.a expressed confidence intervals with correlation coefficients of the minimum confidence limit higher than $\pm$ 0.4.

Moderate correlation coefficients were thus obtained for some crop types (winter wheat, barley, and sugar beets). It is clear that the July 06 acquisition including MSS4 produced the highest correlation coefficients. Various logarithmic applications did not significantly change the relationships.

Analysis of variance between yield classes ((i) lower than mean yield - higher than mean yield, (ii) low - mean - high yield revealed statistically significant differences mainly for cases with statistically significant correlation coefficients. Some examples of analyses of variance plots are given in Figs. 6.6.a-b.

Very low correlation coefficients were obtained for summer wheat, rye, oats, peas, grazing crops and rape seed (Appendix B).

Since the statistical material is weak (too few fields) it is difficult to draw reliable conclusions from the results. It is however a fact that for certain crop types (winter wheat, barley and sugar beets) analyses produced statistically significant relationships between satellite data and yield. Results indicate that it cannot be excluded that LANDSAT data have the capability of providing valuable information to yield prediction models. It is very important to examine the repeatability of these results on a data set expanded in time and space.

6.4.3 Crop biological yield

Correlation coefficients and scatter plots of the relationship between objective yield prediction data and TM data are presented in Appendix C. This appendix concerns each $1m^2$ plot and each crop type within each sub-area and spectral radiance ($L\lambda$).

The cases from Appendix C with minimum confidence limits of the correlation coefficient higher than 0.4 are presented in Tab. 6.6.b.

The crops yielding moderate correlation coefficients were thus winter wheat (area 1 and 2), rye (area 1 and 3) and potatoes (area 3) while the other crops and areas produced extremely low coefficients. It is clear from Tab. 6.6.b that the June 07 acquisition and TM bands 3,5 and 7 or some ratio involving these bands expressed the highest coefficients. The negative relationship between near infrared data and biomass is rather remarkable as is the lack of significant relationships for the IR/R ratio.

Table 6.6.a Pearson's correlation coefficients between crop yield and MSS data.

Crop	n	Band*	Date	Min. conf.	r**	Max. conf.
W.wheat,	46	MSS4/2,	July 06 /	(+0.52) -	(+0.68) -	(+0.84)
Barley,	43	MSS2,	July 06 /	(-0.42) -	(-0.61) -	(-0.79)
Barley,		MSS4/2,	July 06 /	(+0.51) -	(+0.67) -	(+0.83)
Barley,		logMSS2,	July 06 /	(-0.43) -	(-0.62) -	(-0.80)
Sugar,	20	MSS4,	July 06 /	(+0.54) -	(+0.74) -	(+0.94)
beets		logMSS4,	July 06 /	(+0.54) -	(+0.74) -	(+0.94)

* = haze corrected spectral radiance
**= statistically significant (the Student's t test, the 95% level). Williams (1984)

Table 6.6.b Pearson's correlation coefficients between crop biomass and TM data.

Crop	n	Band*	Date	Min. conf.	r**	Max. conf.
area 1						
w.wheat,	29	TM7,	June 07 /	(-0.43) -	(-0.65) -	(-0.86)
		TM7/3	June 07 /	(-0.43) -	(-0.65) -	(-0.86)
area 1						
rye,	23	TM5/3	June 07 /	(-0.43) -	(-0.66) -	(-0.89)
area 2						
w.wheat,	110	TM5	June 07 /	(-0.41) -	(-0.46) -	(-0.51)
		TM7	June 07 /	(-0.43) -	(-0.48) -	(-0.52)
		TM7/3	June 07 /	(-0.43) -	(-0.48) -	(-0.52)
		TM4/7	June 07 /	(+0.43) -	(+0.48) -	(+0.52)
		logTM5	June 07 /	(-0.41) -	(-0.46) -	(-0.51)
		logTM7	June 07 /	(-0.44) -	(-0.44) -	(-0.54)
area 3						
rye,	38	TM3	June 07 /	(-0.42) -	(-0.62) -	(-0.82)
		TM5	June C7 /	(-0.41) -	(-0.61) -	(-0.81)
		TM7	June 07 /	(-0.45) -	(-0.64) -	(-0.83)
area 3						
potatoes,	62	TM3	June 07 /	(-0.53) -	(-0.67) -	(-0.81)
		TM5	June 07 /	(-0.41) -	(-0.57) -	(-0.73)
		TM7	June 07 /	(-0.49) -	(-0.64) -	(-0.79)
		TM4/7	June 07 /	(+0.41) -	(+0.57) -	(+0.73)
		TM4-3/4+3	June 07 /	(+0.45) -	(+0.61) -	(+0.76)
		logTM3	June 07 /	(-0.51) -	(-0.66) -	(-0.51)
		logTM7	June 07 /	(-0.43) -	(-0.59) -	(-0.75)

*= spectral radiance
**= statistically significant (the Student's t test, the 95% level). Williams (1984)

```
ANALYSIS OF VARIANCE
SOURCE       DF         SS          MS          F
FACTOR        2     0.7971      0.3985      17.40
ERROR        43     0.9851      0.0229
TOTAL        45     1.7822
                                           INDIVIDUAL 95 PCT CI'S FOR MEAN
                                           BASED ON POOLED STDEV
LEVEL         N       MEAN       STDEV     --------+---------+---------+--------
37-52        15     2.2083      0.1480     (-----*----)
53-61        15     2.4047      0.1513                   (-----*----)
62-80        16     2.5271      0.1545                               (-----*----)
                                           --------+---------+---------+--------
                                                 2.24      2.38      2.52


ANALYSIS OF VARIANCE
SOURCE       DF         SS          MS          F
FACTOR        1     0.8178      0.8178      37.31
ERROR        44     0.9645      0.0219
TOTAL        45     1.7822
                                           INDIVIDUAL 95 PCT CI'S FOR MEAN
                                           BASED ON POOLED STDEV
LEVEL         N       MEAN       STDEV     --------+---------+---------+--------
37-58        23     2.2499      0.1511     (----*-----)
59-80        23     2.5166      0.1449                               (----*----)
                                           --------+---------+---------+--------
                                                 2.28      2.40      2.52
```

Figure 6.6.a Analysis of variance between crop yield classes of winter wheat. Level= dt/ha, Mean= MSS4/MSS2, July 06.

```
ANALYSIS OF VARIANCE
SOURCE       DF         SS          MS          F
FACTOR        2      2.497       1.248       7.01
ERROR        40      7.127       0.178
TOTAL        42      9.624
                                           INDIVIDUAL 95 PCT CI'S FOR MEAN
                                           BASED ON POOLED STDEV
LEVEL         N       MEAN       STDEV     -----+---------+---------+---------+-
20-34        15     2.1959      0.5295      (------*-------)
35-39        15     2.5029      0.3873                (------*-------)
40-48        13     2.7937      0.3029                          (-------*-------)
                                           -----+---------+---------+---------+-
                                              2.10      2.40      2.70      3.00


ANALYSIS OF VARIANCE
SOURCE       DF         SS          MS          F
FACTOR        1      2.369       2.369      13.39
ERROR        41      7.255       0.177
TOTAL        42      9.624
                                           INDIVIDUAL 95 PCT CI'S FOR MEAN
                                           BASED ON POOLED STDEV
LEVEL         N       MEAN       STDEV     -----+---------+---------+---------+-
20-37        21     2.2435      0.5237      (------*-------)
38-48        22     2.7130      0.2903                          (-------*-------)
                                           -----+---------+---------+---------+-
                                              2.16      2.40      2.64      2.88
```

Figure 6.6.b Analysis of variance between crop yield classes of barley. Level=dt/ha, Mean= MSS4/MSS2, July 06.

Table 6.7 Statistically significant differences between crop biological yield classes.

Crop number:	AREA 2 1	2	3	4	5	6	7	AREA 3 1	2	3	4	5	6	7	AREA 5 1	2	3	4	5	6	7
JUNE 07																					
TM3	Y	N	N	N	N	N	Y	Y	N	N	N	N	N	Y	Y	Y	N	N	N	N	N
TM4	Y	N	N	Y	Y	N	Y	Y	N	N	N	Y	N	Y	N	N	N	N	N	N	N
TM5	Y	N	N	N	N	N	Y	Y	N	N	Y	N	N	Y	Y	N	N	N	N	N	N
TM7	Y	N	N	N	N	N	Y	Y	N	N	N	N	N	Y	Y	N	N	N	N	N	N
TM4/3	Y	N	N	Y	Y	N	Y	Y	N	N	N	Y	N	Y	Y	N	N	N	N	N	N
TM7/3	Y	N	N	N	N	N	Y	Y	N	N	Y	N	N	Y	Y	N	N	N	N	N	N
TM4/7	Y	N	N	Y	N	N	Y	Y	N	N	N	N	N	Y	Y	N	N	N	N	N	N
TM5/3	N	N	N	N	N	N	Y	Y	N	N	N	N	N	N	N	N	N	N	N	N	N
ND	Y	N	N	Y	Y	N	Y	N	N	N	N	Y	N	Y	Y	N	N	N	N	N	N
JULY 09																					
TM3	N	N	N	N	N	N	N	N	N	N	N	Y	N	N	N	N	N	Y	N	N	N
TM4	Y	N	N	Y	Y	N	N	Y	N	N	N	Y	N	N	Y	N	N	Y	N	N	N
TM5	Y	N	N	N	N	N	N	Y	N	N	N	Y	N	N	Y	N	Y	Y	Y	N	N
TM7	Y	N	N	N	N	N	N	Y	N	N	N	Y	N	N	Y	N	Y	Y	Y	N	N
TM4/3	Y	N	N	Y	Y	N	N	Y	N	N	N	Y	N	N	Y	N	N	Y	N	N	N
TM7/3	Y	N	N	N	Y	N	N	Y	N	N	N	Y	N	N	Y	N	N	Y	N	N	N
TM4/7	Y	N	N	N	N	N	N	Y	N	N	N	Y	N	N	Y	N	N	Y	N	N	N
TM5/3	Y	N	N	N	N	N	N	Y	N	N	N	Y	N	N	N	N	N	Y	N	N	N
ND	Y	N	N	Y	Y	N	N	Y	N	N	N	Y	N	N	N	N	N	Y	N	N	N

1= Winter wheat 5= Oats
2= Summer wheat 6= Potatoes
3= Rye 7= Grazing crop
4= Barley
Y= Statistically significant differences (the 95% level) between satellite data of the two biomass classes lower than mean biomass and higher than mean biomass
N= Not statistically significant differences (the 95% level) between satellite data of the two biomass classes lower than mean biomass and higher than mean biomass.

Application of various:

- logaritmic functions

- options (i) each $1m^2$ plot in each sub-area, (ii) each $1m^2$ plot in all sub-areas together, (iii) the mean of two $1m^2$ plots in each sub-area and (iv) the mean of two $1m^2$ plots in all sub-areas together

- preprocessing alternatives

did not significantly change correlation coefficients.

Appendix C also reveals that the amount of biomass (g/m^2) of the two $1m^2$ plots belonging to the same field was connected to each other and that the variation of the TM4/TM3 ratio with biomass was not convincing.

For some area/crop/band/combinations, statistically significant differences were obtained between the biomass classes lower than mean biomass and higher than mean biomass (Tab. 6.7). Winter wheat was the only crop producing statistically significant differences in all individual sub-areas and in all sub-areas together.

An indication of a relationship between spectral data and biological yield was thus obtained for winter wheat. Generally, however, the relationship was weak. It must be concluded that the type of data used in this work (i.e. the pixel overlaying the coordinated $1m^2$ plot) is not suitable for integration with the objective yield prediction procedures used today. It should be noted that this study <u>does not</u> exclude a relationship between biological biomass and satellite measurements regarding other types of field data.

6.5 STUDY REMARKS

* Significant radiometric errors may have been introduced as a result of resampling (only the TM data).

* The accuracy in space between a $1m^2$ test plot and the TM pixel "overlaying" this plot was not further considered.

* The relationship between spectral data and LAI was not examined.

* Too few cases particularly for the MSS data (the author is very well aware of this limitation but the collection of yield data proved to be a troublesome and extremely time-consuming task).

* Since the $1m^2$ is not representative of the whole TM pixel, the absolute relationship between biological biomass and

satellite data could not be examined.

* Only half of the radiance value for each pixel may be related to the ground resolution cell it represents (Curran and Hay 1986). This is caused by the point spread function of the sensor, velocity smear, and atmospheric scattering.

* The temporal-spectral profiles were strongly influenced by solar elevation correction approach (no correction, $\rho P\lambda$ or $B\lambda,s$). Since ground-based measurements could not be accomplished, it was not possible to evaluate these approaches.

* Lord et al. (1985a) and Parsons (1985) found a large influence of wind on crop canopy spectral reflectances collected from ground-based platforms. Possible wind effects were not considered, nor were possible row effects (Suits 1983).

* Crop variety variation within each crop type was not regarded.

* As the solar spectral irradiance differes between various bands a ratio between exoatmospheric reflectance is more adequate than a ratio between spectral radiance.

6.6 SUMMARY

The aim of this section was to:

1. Study the relationship between crop type and satellite data (MSS and TM data).

2. Study the relationship between crop yield and satellite data (MSS data).

3. Achieve a very preliminary estimation of the value of integrating satellite measurements in the objective yield prediction performed in Sweden each year.

Results:

Point 1

- Mono- and multitemporal studies produced statistically significant differences between winter and summer crops for all acquisition dates as well as temporal-spectral profiles with varying trends.

- Rape seed, potatoes and sugar beets differed significantly from the other crops on all acquisition dates.

- Individual cereals seldom produced statistically significant differences. In July, however, rye differed statistically from summer wheat, barley and oats.

It was concluded that a promising possibility exists concerning classification potentials of crop type groups and certain individual crop types.

Point 2

- Statistically significant correlation coefficients were obtained between yield of certain crop types (winter wheat, barley, and sugar beets) and satellite data. The importance of analysing the repeatability of these results was underlined.

Point 3

- A weak relationship was obtained between objective yield prediction data and TM data.

7

CROP CLASSIFICATION USING LANDSAT TM DATA

In section 6 studies of the relationship between crop type and satellite data are presented. Statistically significant differences between crop type and satellite data were frequently obtained. It was concluded that a promising possibility exists concerning classification potentials of crop type groups as well as of certain individual crop types.

The basic aim of the studies was to:

- calculate the classification accuracy of some common Swedish agricultural crops

- analyse the relationship between actual and classified field size.

Additionally:

- various methods to calculate classification accuracy value were compared

- the influence of field size on classification accuracy was studied

- classification results of data treated with various pre-processing alternatives were compared

- the possible influence of soil background on classification results was analysed.

7.1 BACKGROUND

7.1.1 Classification methods

Two main classification approaches exist: (i) supervised classification and (ii) unsupervised classification.

In <u>unsupervised</u> methods the pixels are grouped into classes automatically (clustering), without any other given input than the number of classes.

The use of <u>supervised</u> methods requires preliminary knowledge about the classes (i.e. training from areas with known ground data). Supervised methods may be separated into:

(i) classical (also called per-pixel and noncontextual) methods

(ii) contextual methods (Saebö 1985).

In classical approaches (so far the most frequently used) each pixel is classified on the basis of the data of only this pixel. This single-pixel process does not compete with human abilities, which are clearly distinguished by their use of texture and context in addition to tone to aid interpretation decisions (Harris 1985). The maximum likelihood (ML) method is frequently used in per-pixel algorithms.

Data from neighbouring pixels are considered in the contextual approach. Thus, information concerning how the classes vary in space and are located to each other is included in the contextual classification procedure.

7.1.2 Review of literature

- Unsupervised classification.

Justice and Townshend (1982) compared two unsupervised classification procedures applied to ratioed and unratioed LANDSAT MSS data. Classification of the unratioed data produced higher percentage accuracies than the ratioed data. A completely automated, unsupervised classification approach was developed by Badhwar (1984a) and applied to classification of Argentine summer crops by Badhwar et al. (1987).

- Comparison between unsupervised and supervised classification.

For a multicrop test area Csillag (1986) compared two supervised and two unsupervised classification approaches ((i) interpreter trained ML, (ii) interpreter trained polygon-vector, (iii) ISODATA-type clustering with Euclidean intercluster distance, (iv) ISODATA-type clustering with Swain-Fu intercluster distance). The best results were obtained by clustering using the Swain-Fu distance (unsupervised approach). Interpreter trained ML and polygon vector classification (supervised approaches) generally led to decreased accuracy. Comparison of unsupervised and supervised methods was also performed by Lo et al. (1986) and comparable results were found in this study. Thus the best results were produced by unsupervised classification.

- Supervised - classical classification.

Batista et al. (1985) used a per point Gaussian ML classifier to identify corn and soy-bean. Mather (1985) developed a computationally-efficient ML classifier. A fruit tree inventory based on the ML approach was performed by Gordon et al. (1986). Moreira et al. (1986) used an ML classifier and a post-classification spatial filtering to classify wheat. A study of the effect of spatial resolution and degree of internal variability within land-cover types using an ML classification was accomplished by Cushnie (1987).

- Supervised - contextual classification.

Megier and Mehl (1984) used a classification method, based on global statistics defined over individual agricultural fields. Compared with a pixel by pixel approach this method increased the classification accuracy around 18%. A contextual post-classification procedure was applied to improve ML classification in arid and urban areas in Kuwait by Harris (1985).

Classification of particularly agricultural crops was performed by e.g. Badhwar (1984a,b,c,d), Megier and Mehl (1984), Batista et al. (1985), Csillag (1986), Koo and Kim (1986), Lo et al. (1986), Moreira et al. (1986), Badhwar et al. (1987).

Megier and Mehl (1984) studied the possibility to classify e.g. vine, winter wheat, summer wheat, winter barley, spring barley, lucerne, melons, sugar beets, vegetables, maize, oats in a flat agricultural area in FRG. Various classification approaches applied on SPOT simulated data and SAR data were used. The best result was a 90% classification accuracy for five agricultural categories.

Batista et al. (1985) classified corn and soybean in the Corn Belt region (United States). The overall accuracy was 83.9%. The authors stress the influence of field size on classification results, with small fields tending to have low accuracies even when the effect of mixed pixels is eliminated.

Fields of wheat, corn, alfalfa, forest, sunflower, grass and water surfaces (Hungary) were classified by Csillag (1986). Accuracy was measured on a per-point basis of the proportion of correctly classified pixels. The best accuracy (95%) was obtained using an unsupervised approach while the "traditional" supervised ML method produced 80% accuracy.

Lo et al. (1986) classified corn, alfalfa, oats and six natural surfaces (Wisconsin, United States). The accuracy, weighted by the number of pixels occurring in each class, was (i) corn 97.3%, (ii) alfalfa 87.5% and (iii) oats 79.5%. The overall accuracy for all nine classes was 89.8%.

In Argentina, Badhwar et al. (1987) classified 75 homogeneous fields (21 soybean, 12 sunflower, 8 sorghum, 9 corn, 25 pasture) and in addition 145 fields of winter wheat, oats, pastures, and bare soil. The average field size was about 150ha. The classification accuracy of individual crops using an ML method in which the classifier was trained in the fields and the same fields classified ranged from 80% to 100%.

Several successful classification studies include application of tasseled cap transformations (Badhwar 1984a,b,c,d, Lo et al. 1986, Csillag 1986, Badhwar et al. 1987). The tasseled cap transformations were developed in the U.S. Corn Belt. These transformations have been used in other regions by Csillag (1986) (Hungary), Badhwar et al. (1987) (Argentina) and Andersson and Gervais (1987) (Sweden - soil mapping). Badhwar et al. (1987) maintain that the soil line developed in the United States areas can be extended to areas in Argentina and they thus propose that the transformations are applicable to foreign countries.

7.2 MATERIAL AND METHODS

7.2.1 Material

Remote sensing data included LANDSAT TM acquisitions of the province of Östergötland (1984; May 19, June 20 and July 06, section 2).

Crop data (including the size in hectares of each field) were obtained through cooperation with the Central Bureau of Statistics (SCB), (section 6.3.1 and Tab. 6.1.b) and comprised winter wheat, summer wheat, rye, barley, oats, potatoes and grazing crop.

7.2.2 Methods

7.2.2.1 Training

In the classification of TM data, randomly selected 1m² plots (section 6.3.1) were used to characterize the seven classes considered. The boundaries of the fields to which these plots belonged were easily determined on A DIPIX ARIES screen. Therefore, all pixels belonging to the field were included in the training. These fields were spread in all six sub-areas (Fig. 4.1). The number of training pixels for each class was:

winter wheat - 784
summer wheat - 535
rye - 646
barley - 1104
oats - 966
potatoes - 598
grazing crop - 645

7.2.2.2 Classification approach

Using the maximum likelihood algorithm on the DIPIX ARIES III image processing system, classified images were produced. This algorithm works by first calculating the mean vector variance and correlation for each land cover class in the training data, on the assumption that the data for each class are normally distributed. The spread of pixels around each mean vector is described by a probability function and pixels from the whole data set are allocated to the class with which they have the highest probability of membership. To improve the accuracy of the classifier, a post classification "clean-up" filter was used.

Table 7.1 The preprocessing alternatives applied to the maximum likelihood classification.

BASIC DATA																			TRANS-FORMA-TION	NO. OF BANDS
	June 07						July 09						August 26							
TM	1	2	3	4	5	7	1	2	3	4	5	7	1	2	3	4	5	7		
1.							X	X	X	X	X	X							none (DN)	6
2.							X	X	X	X	X	X							$L\lambda$	6
3.							X	X	X	X	X	X							$p\rho\lambda$	6
4.							X	X	X	X	X	X							$p\rho\lambda$, standardized,	6
5.							X	X	X	X	X	X							$B\lambda,s$	6
6.							X	X	X	X	X	X							*PCA of $p\rho\lambda$	6
7.			X	X					X	X					X	X			$p\rho\lambda$	6
8.			X	X					X	X					X	X			$p\rho\lambda$, standardized,	6
9.			X	X					X	X					X	X			$B\lambda,s$	6
10.			X	X					X	X					X	X			TM4/3 of $p\rho$	3
11.			X	X					X	X					X	X			*PCA of DN,	(*1)
12.			X	X					X	X					X	X			*PCA of $p\rho$,	(*1)
13.	X	X	X	X	X	X	X	X	X	X	X	X	X	X	X	X	X	X	*PCA of $p\rho$,	(*2)
14.	X	X	X	X	X	X	X	X	X	X	X	X	X	X	X	X	X	X	*PCA of DN,	(*3)
15.	X	X	X	X	X	X	X	X	X	X	X	X	X	X	X	X	X	X	*PCA of $p\rho$,	(*3)

1-6= monotemporal classification
7-15= multitemporal classification
*= only subarea 2 (Fig. 4.1)
*1= first 4 components
*2= first 3 components of each date
*3= each 2nd component of each date

Classification was performed on original data (DN) and on transformed data ($L\lambda$, $p\rho\lambda$, $B\lambda,s$, standardization - section 4, principal component analysis etc.). Tab. 7.1 presents the transformations applied to the ML classifier.

Principal component analysis (PCA) is a data transformation technique commonly used in multivariate statistical analysis. The aim of this technique in remote sensing is to remove redundancy in the multispectral image by rotating the axes into principal components which maximize variance in the data. The first component will contain the largest possible amount of total variance, the second component will contain the largest possible amount of the remaining variance, and so on.

The tasseled cap (Kauth and Thomas 1976), the two dimensional perpendicular vegetation index and the soil line index (Richardson and Wiegand 1977, Wiegand and Richardson 1982), are transformations similar to the principal component transformation. The main difference is that these transformations are based on a prior knowledge of the spectral behaviour of soil and vegetation, while this is not the case for PCA.

7.2.2.3 Accuracy estimation

The most common way to express the accuracy of a classified image (map) is by a declaration of the percentage of the map area that has been correctly classified when compared with reference data. An error matrix (confusion matrix or contingency table) is usually the base for such a declaration. In the matrix the reference data (represented by the columns of the matrix) are compared to the classified data (represented by the rows). The major diagonal expresses the agreement between these two data sets and from the matrix it is possible to calculate the classification accuracy for all classes together as well as the accuracy for each individual class. In the error matrices of the present study, the reference data consisted of the pixel overlaying the $1m^2$ plot with known crop type (sections 6.3.1-2) and the classified data the classification result of this pixel.

The error matrix enables several ways to calculate classification accuracy. In the present study various methods presented in literature have been used and compared:

1. Producer's accuracy (Hellden 1980, Story 1986)
2. User's accuracy (Hellden 1980, Story 1986)
3. Mean accuracy (Hellden 1980)
4. Overall accuracy (Hellden 1980, Story 1986)
5. Kappa coefficient of agreement as a measure of total map accuracy (Rosenfield and Fitzpatric-Lins 1986)
6. Conditional Kappa as a measure of individual category

accuracy (Rosenfield and Fitzpatric-Lins 1986).

These approaches are presented below.

1. Interpretation accuracy (Hellden 1980) or producer's accuracy (Story 1986),

   $(X_{ii}/X_{+i})*100(\%)$ (7.1)

   where

   X_{ii}= number of correctly mapped sample units for a specific class
   X_{+i}= number of sample field units for a specific class
   X_{i+}= number of sampled map units for a specific class

   The number of correctly classified samples of category X is divided by the total number of reference samples of this category. The resulting percentage accuracy indicates the probability that a reference (ground) sample will be correctly classified and is in reality a measure of errors of omission (samples that are incorrectly classified are omitted from the correct category). According to Story (1986), this accuracy value is the most common and widely accepted and since the producer of a classified map is interested in how well a specific area on the Earth can be mapped it may be referred to as "producer's" accuracy.

2. Object accuracy (Hellden 1980) or user's accuracy (Story 1986),

   $(X_{ii}/X_{i+})*100(\%)$ (7.2)

   The number of correctly classified samples of category X is divided by the total number of samples that were classified in this category. The resulting percentage accuracy indicates the probability that a sample from the classified image actually represents that category on the ground and is in reality a measure of errors of commission (Story 1986). Since a map user is interested in how well the map represents what really is on the ground, this accuracy value may be referred to as "user's" accuracy (Story 1986).

Story (1986) points out that a single accuracy value (producer's or user's) can be extremely misleading and that too often only one measure of accuracy is reported. The author emphasizes the importance of always considering producer's as well as user's values when reporting classification accuracies.

3. Mean accuracy (Hellden 1980),

$(2X_{ii}/(X_{+i}+X_{i+}))*100(\%)$ (7.3)

Mean accuracy denotes the probability that a randomly chosen point of a specific class on the map has a correspondence of the same class in the same position in the field and that a randomly chosen point in the field of the same class has a correspondence of the same class in the same position on the map (Hellden 1980).

4. Total accuracy (Hellden 1980) or overall accuracy (Story 1986),

   $\Sigma X_{ii}/N*100(\%)$ (7.4)

   where N= total number of control points

   This accuracy value denotes the probability that a randomly chosen point on the map has a correspondence of the same class and the same position in the field and that a randomly chosen point in the field has a corresondence of the same class and in the same position on the map.

To test mapping reliability (of producer's, user's, mean, and overall accuracy) the following formula was used (Hellden 1980, Curran 1985):

$SE= SQR((p\% * q\%)/N)$ (7.5)

where

SE= standard error in an assumed distribution of means
p = per cent correct classification
q = 100-p

The p% correct observation was used as a mean with +- 1.96*SE as 95% confidence limits.

Significant differences of sample means were tested using (Hellden 1980):

$Z= (M1-M2)/SQR(S1^2/N1+S2^2/N2)$ (7.6)

where

M1= mean of sample 1, S1= standard deviation of sample 1
M2= mean of sample 2, S2= standard deviation of sample 2
N1= number of points in sample 1
N2= number of points in sample 2

Differences were considered to be significant at the 95% confidence level when Z was smaller than -1.95 or larger than +1.95 (Hellden 1980).

So far methods of accuracy estimation have only concerned errors by commission and errors of omission (Hellden 1980, Story 1986). An alternative way is to apply the problem as a measure of agreement between classification and reference (Rosenfield and Fitzpatric-Lins 1986). The Cohen's Kappa coefficient of agreement is a measure of agreement that has attracted attention by remote sensing workers (Congalton and Mead 1983, Rosenfield and Fitzpatric-Lins 1986).

5. Kappa coefficient of an agreement as a measure of total map accuracy (Congalton and Mead 1983, Rosenfield and Fitzpatric-Lins 1986) is calculated by

$$\hat{K} = \frac{\sum_{i=1}^{r} X_{ii} - \sum_{i=1}^{r} (X_{i+} \; X_{+i})}{N^2 - \sum_{i=1}^{r} (X_{i+} \; X_{+i})} \qquad (7.7)$$

where

r = number of rows and columns in error matrix
X_{ii} = number of observations in row i and column i,
X_{i+} = marginal total of row i,
X_{+i} = marginal total of column i,
N = total number of observations

This measure of agreement uses all cells in the matrix, not just diagonal elements and constitutes the proportion of agreement after chance agreement is removed from consideration. When K=0 the obtained agreement equals chance agreement. The upper limit of Kappa (+1.00) occurs only when agreement is perfect. In the present study Kappa coefficients were multiplied by 100 (Rosenfield and Fitzpatric-Lins 1986).

6. The coefficient of conditional agreement (conditional Kappa) as a measure of individual class accuracy (Rosenfield and Fitzpatric-Lins 1986) is calculated by

$$\hat{K}_i = \frac{N X_{ii} - X_{i+} X_{+i}}{N X_{i+} - X_{i+} X_{+i}} \qquad (7.8)$$

The Kappa coefficients use information in the error matrix that results from errors by commission and omission and the Kappa method has the statistical basis of being an intraclass correlation coefficient. Therefore, Rosenfield and Fitzpatric-Lins (1986) recommended that coefficients of Kappa and conditional Kappa are adopted by the remote sensing community as a measure of accuracy for thematic classification as a whole as well as for individual categories.

Detailed presentations and practical applications of the Kappa coefficient of agreement are given by Bishop et al. (1975), Congalton and Mead (1983), Rosenfield and Fitzpatric-Lins (1986). Methods for calculating large sample variance and confidence limits of Kappa are given by Bishop et al. (1975) and methods for testing the significance between two different Kappas by Rosenfield and Fitzpatric-Lins (1986). Some formulas and numerical results presented in the literature (Bishop et al. 1975, Congalton and Mead 1983, Fritzpatric-Lins 1986) are unfortunately incorrect. Accurate formulas and correct numerical results are given by Hudson and Ramm (1987). These authors point out that the Kappa coefficient of agreement is very useful in remote sensing accuracy assessment but that the user of the Kappa coefficient of agreement should be conscious of its correct formulation and numerical computation.

In the literature, some discussion has been devoted to how many (if any) of the training pixels may be used for accuracy estimation, and how these pixels should be included in the estimation. Yool et al. (1986) and Badhwar et al. (1987) performed a back-classification, i.e. used the same sites for testing as for training. In the present study training pixels were not included in accuracy estimation.

7.2.2.4 Integration of soil map data and satellite data

Several authors have demonstrated that the spectral response of vegetation canopies may vary with different soil backgrounds (Elvidge and Lyon 1985, Huete et al. 1985, Gardner and Blad 1986, Heilman and Boyd 1986).

In order to study to what extent the obtained classification accuracies may be influenced by soil type variations, the relationship between satellite data of various soil types was analysed. Thus, the soil type of each m^2 plot with known crop type (section 6.3.1-2) was identified through soil maps (Swedish Geological Survey (SGU), series Ae, 1:50 000). For each crop type, analyses of variance were performed between the pixel values (overlaying each m^2 plot) of various soil types. Analysis of variance included the F-test at the 95% confidence level (Ryan et al. 1985).

The following soil types (based on the soil maps of SGU) were identified: thin peat cover, postglacial clay, postglacial silty clay, postglacial coarse silt, fine sand, sand, varved clay, glaciofluvial deposit - unspecified, till - sandy.

7.2.2.5 Procedure

Based on the procedure summarized in Tab. 7.2 the following items were considered:

- The classification accuracy for all classes together - all sub-areas together (Tab. 6.1.b).
- The classification accuracy for all classes together - each sub-area.
- The classification accuracy of individual classes - each sub-area.
- Differences between various methods to calculate classification accuracy values (points 1-6 section 7.2.2.3)
- Statistically significant differences between various transformations (points 1-15 Tab. 7.1).
- The relationship between actual and classified field size.
- The relationship between field size and classification accuracy.
- Possible influence of soil type variations on classification accuracy.

<u>Table 7.2</u> Summation of the procedure applied.

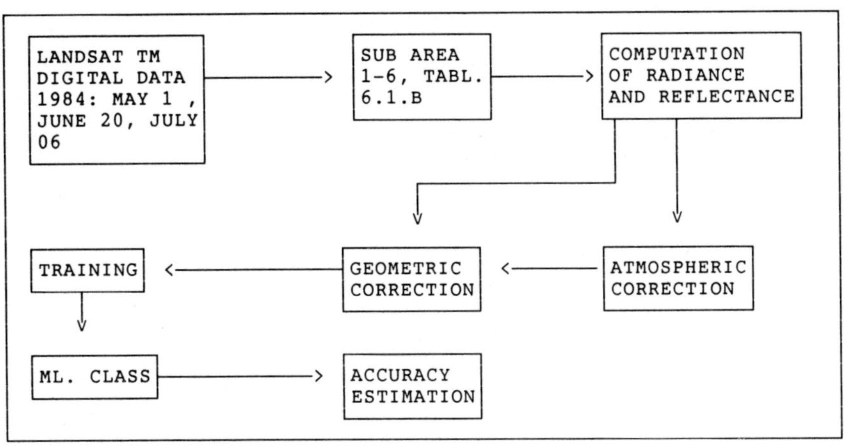

7.3 RESULTS

7.3.1 Accuracy estimation

7.3.1.1 Total map accuracy

The classification accuracy of all classes together, calculated as Kappa coefficient of agreement (Rosenfield and Fitzpatric-Lins 1986) as well as overall accuracy (Hellden 1980), for all sub-areas together (Tab. 6.1.b) and for each transformation (Tab. 7.1) are presented in Figs. 7.1.a-b. The Kappa coefficients (Fig. 7.1.a) were commonly lower than the overall accuracies (Fig. 7.1.b) and were also associated with broader confidence limits. The Kappa coefficient of agreement yielded values around 55% while overall accuracies were around 65%. Compared with Batista et al. (1985) and Csillag (1986) (section 7.1.2), the present study produced lower overall accuracy values.

For Kappa values statistically significant differences were obtained between transformation 14 and 9, with 14 producing the highest accuracy. The reliability of this result is limited however, since transformation 14 is based on only sub-area 2. Statistically significant differences were obtained between transformation 14 and all other transformations for overall accuracies. Other transformations did not express statistically significant differences.

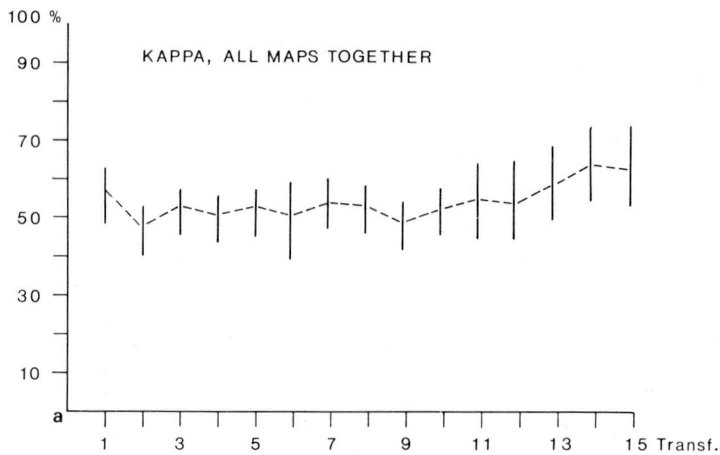

Figure 7.1.a The Kappa coefficient of agreement as a measure of total map accuracy. For all sub-area together. Mean and 95% confidence intervals.

The Kappa coefficient of agreement as a measure of total map accuracy and overall accuracy for the individual sub-areas 1-6 are presented in Appendix D. As can be seen in the table, various sub-areas expressed roughly similar accuracies. Approximate differences between sub-areas were between 5-15% with sub-area 4 often yielding the lowest accuracies (in a few cases statistically significant differences at the 95% level).

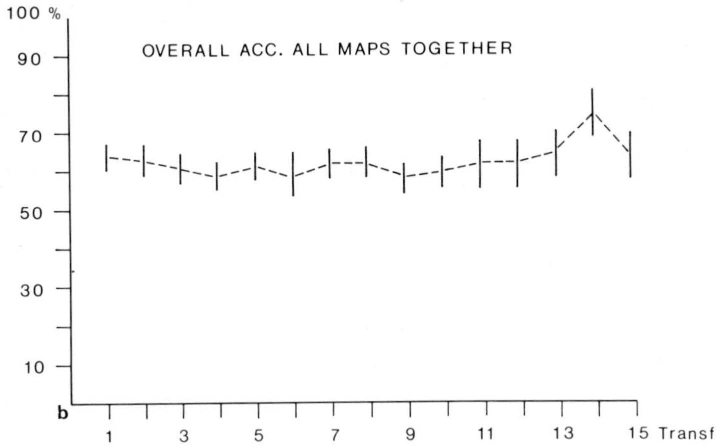

Figure 7.1.b The overall accuracy for all sub-areas together. Mean and 95% confidence intervals.

7.3.1.2. Individual crop accuracy

Some examples of producer's accuracy, user's accuracy, mean accuracy and conditional Kappa together with 95% confidence limits for each class within each sub-area (Fig. 4.1) and for each transformation (Tab. 7.1) are given in Figs. 7.2.a-g, Appendix E and Appendix F. Figs. 7.2.a-g illustrate the Kappa accuracies with 95% confidence limits of sub-area 2, for individual crops and transformations. The Kappa, mean, producer's and user's accuracies with 95% confidence limits for individual crops, sub-areas 1-6 and transformations 3-4 are tabulated in Appendix E. Appendix F gives some mixed examples of accuracy values of various accuracy estimation methods, transformations and sub-areas. Appendix G constitutes an effort to summarize the results of the individual crop classification. However, as the reader carefully studying Figs. 7.2.a-g, and Appendices E,F will notice, the accuracy values vary extremely for various sub-areas. Therefore, Appendix G represents merely a strong generalization of results.

A general impression of the individual crop classification

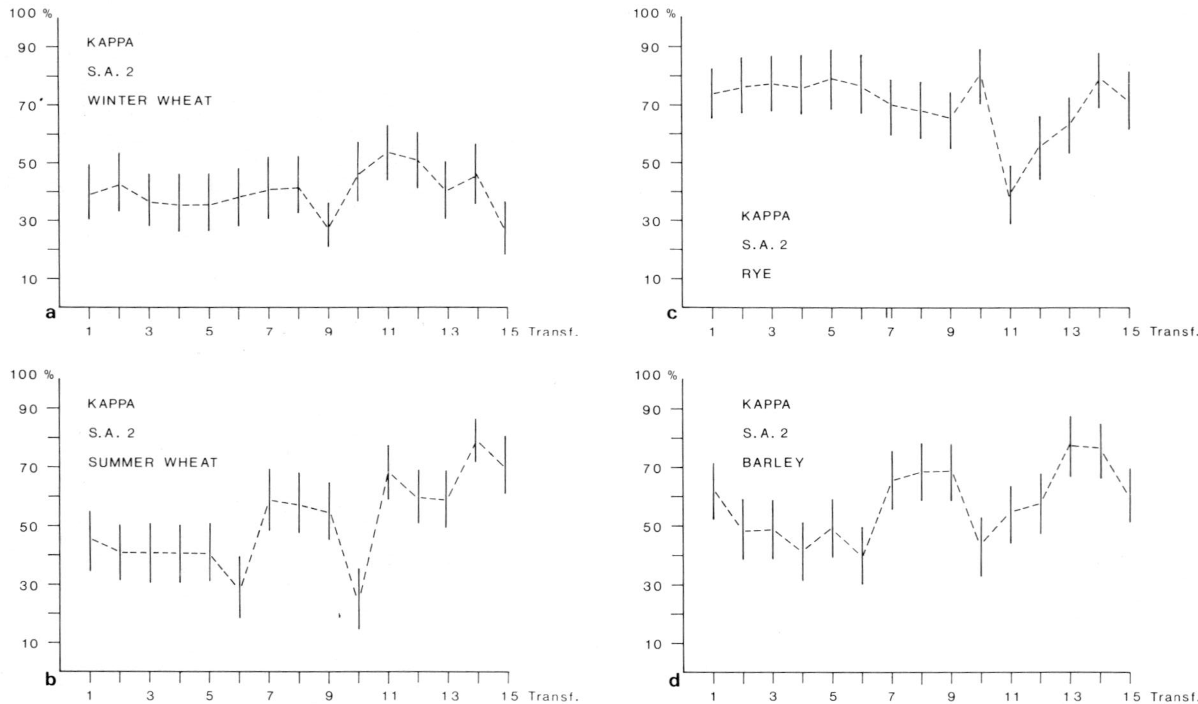

<u>Figure 7.2</u> The conditional Kappa accuracies with 95% confidence intervals for transformations 1-15 (Tab. 7.1).

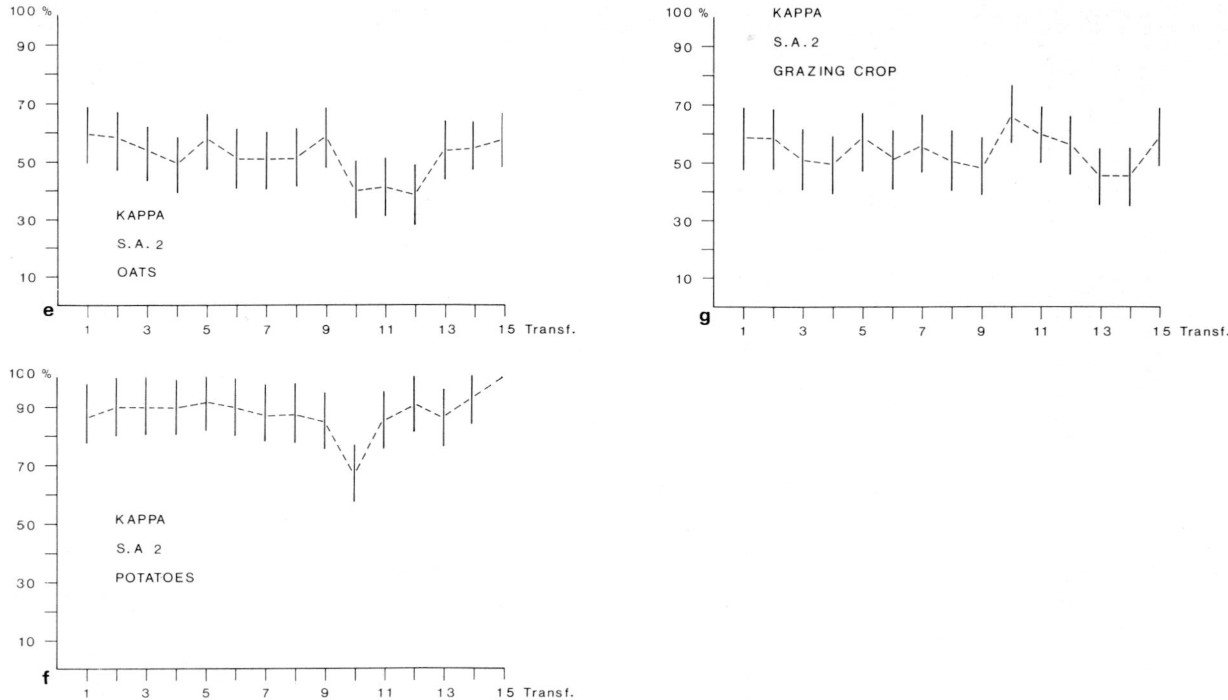
100 %
90
70
50
30
10
KAPPA
S.A. 2
OATS
e
1
3
5
7
9
11
13
15 Transf.
100 %
90
70
50
30
10
KAPPA
S.A 2
POTATOES
f
1
3
5
7
9
11
13
15 Transf.
100 %
90
70
50
30
10
KAPPA
S.A.2
GRAZING CROP
g
1
3
5
7
9
11
13
15 Transf.

Fig. 7.2 cont.

is that rye and potatoes produced the highest accuracy values. Kappa coefficients for rye were roughly around 75% and for potatoes around 70% while mean, producer's and user's accuracies were around 60-65%. The other crops (winter wheat, summer wheat, barley, oats and grazing crops) often produced lower accuracy values with Kappa values for winter and summer wheat around 40% and for barley, oats and grazing crop around 55%. Mean accuracy value for these crops was 50-60% while producer's accuracy was around 60% (Figs. 7.2.a-g, Appendices E,F,G). Badhwar et al. (1987) classified sorghum, corn, winter wheat and oats and obtained individual accuracy values between 80% and 100%. Compared with classification accuracies presented in the literature the present study thus produced low accuracy values.

Curran (1985) states that the classification accurcay is negatively related to the number of classes. This was noticed in the present study as well. Thus, when reducing the number of classes from seven to four by joining winter crops on one hand and summer crops on the other hand and keeping potatoes and grazing crops as individual classes increased accuracy values were obtained. The increase was roughly around 10%.

7.3.1.3 Various transformations

Analysis of difference between classification accuracies of various transformations often revealed statistically significant differences. It was however not possible to distinguish certain transformations as always being superior to others. Transformation 1 (DN of a single date, July 09) often produced low accuracies. This was also a fact for transformations 5 and 9. High accuracies were frequently obtained for transformations 3 and 4 (spectral reflectance and standardized spectral reflectance, single date - July 09) as for transformations 13-15 (PCA, multitemporal). Thus, it was not unambiguously proved that a multitemporal classification was more suitable than a monotemporal one.

7.3.1.4 Various accuracy estimation methods

The studied accuracy estimation methods were found to produce similar trends i.e. high Kappa total map accuracies often corresponded to high overall accuracies, high conditional Kappas with high producer's accuracies etc. The absolute magnitudes of various accuracy values varied however. The accuracy methods based on only the diagonal cells of the error matrix produced higher values than the method using information in the error matrix resulting from errors by commission and omission (defensible as intraclass coerrelation coefficients - Kappa coefficients of agreement).

7.3.2 Field size

The mean difference between actual and classified field size for all fields of all crops together were 2.5 ha. 60% of the fields were "overclassified", e.g. the classified area was larger than the actual field size. Generally, a high correspondence between actual and classified field existed. Some typical results are presented in Figs. 7.3.a-c. Fig. 7.3.a gives the plot between actual and classified field size for winter wheat (transformation 14, sub-area 2). The Pearson product-moment correlation coefficient for this plot was 0.883 (95% conf. interv. 0.822-0.944), the mean actual field size was 10.9 while the mean classified field size was 11.4. Fig. 7.3.b shows an example for rye (r= 0.967 (0.944-0.990), the mean actual field size - 9.1 ha, the mean classified field size - 10.7 ha) while Fig. 7.3.c presents an example for potatoes (r= 0.978 (0.954-1.0), the mean actual field size - 4.6 ha, the mean classified field size - 5.0).

The possible influence of field size on classification results was performed by analysis of variance between the size of fields associated with correctly and uncorrectly classified pixels. Uncorrectly classified pixels were coupled with a smaller mean field size than the correctly classified pixels. Generally, these differences were statistically significant. Results thus revealed that the field size may influence the classifiaction accuracy, which in this study was calculated on a single pixel. Analysis of variance between field size of various crop types did not yield statistically significant differences. The varying classification accuracy between various crop types may thus not be explained by differences in field sizes.

7.3.3 Influence by soil background

For each crop type of each sub-area as well as for all sub-areas together, analyses of variance between pixel values of various soil types were performed. This was carried out for all acquisition dates. In several cases statistically significant differences between pixel values of various soil types were obtained. However, when comparing results based on different sub-areas no common trend could be identified, i.e. results based on one sub-area was not repeatable in another sub-area. This means that for one sub-area certain soil types could produce statistically significant pixel value differences while the same type of soil and crop configuration in another sub-area did not produce statistically significant pixel values. Thus, results did not indicate a common unambiguous influence of soil type on pixel values and it was concluded that the classification results were not influenced by soil background. However, since the soil data used were extremely general (section 7.4), these results and conclusions are rather coarse, solely indicating a non-apparent influence. It cannot be

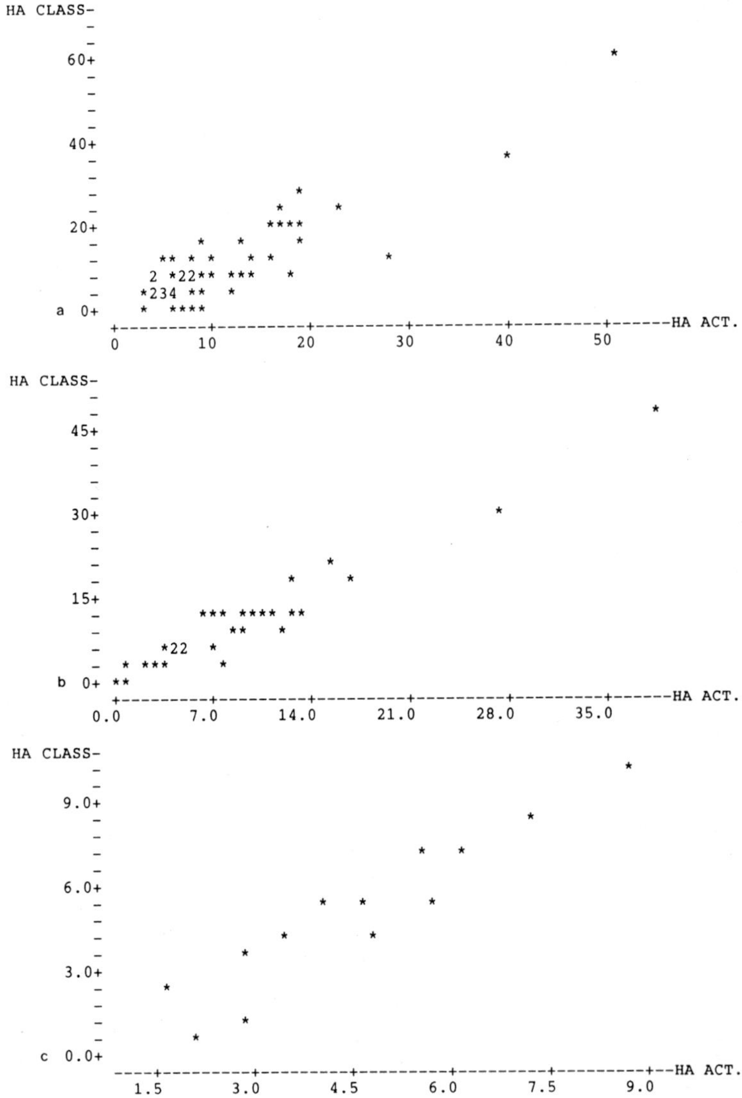

Figure 7.3 The relationship between actual (ACT) and classified (CLASS) field size. HA= hectares.
(a)= winter wheat, r=0.883, n=50
(b)= rye, r=0.967, n=30
(c)= potatoes, r=0.978, n=13
(Valid for sub-area 2, information concerning field size was not available for all fields)

excluded that very detailed soil data (e.g. through field work) would produce different conclusions.

7.4 STUDY REMARKS

* Due to the resampling, radiometric errors may have been inducted.

* Errors may have arisen due to geometric disturbances between ground data and satellite data.

* The most significant function that influences the classification accuracy is training (Swain and Davis 1978). The relationship between training data and classification accuracy should thus be examined. This was not performed in the present work.

* Only one classification algorithm (ML) was used.

* The soil map (1:50 000) produced very general soil type information. Soil type data based on field work would have been superior.

* No spatial atmospheric correction was performed.

7.5 SUMMARY

The objective of the work reported in this section was to (i) calculate the classification accuracy of some common Swedish agricultural crops, (ii) analyse the relationship between actual and classified field size, (iii) analyse the possible influence of field size on classification results, (iv) compare various methods to calculate classification accuracy values, (v) compare classification results of data treated with various preprocessing alternatives, and (vi) analyse the possible influence of soil background on classification results.

The Kappa coefficient of agreement as a measure of total map accuracy yielded accuracy values around 55%. The highest individual Kappa accuracies were approximately 75% (rye and potatoes) while the main part of the individual crops produced accuracies between 50% and 60%. Thus, classification results were not very encouraging and generally the obtained classification accuracy values were somewhat lower than the accuracy values presented in the literature (Batista et al. 1985, Csillag 1986, Badhwar et al. 1987).

The relationship between actual and classified field size was good (correlation coefficients around 0.9).

When the number of classes were reduced from seven to four, increased classification accuracies were obtained.

Various methods to calculate classification accuracies produced similar trends but different values. Methods using only the diagonal elements of the error matrix were found to yield inflated accuracy values compared with the Kappa coefficients of agreement. The importance of always stating the accuracy method used and the need for the remote sensing community to become united concerning adoption of suitable accuracy methods have been highlighted.

Classification accuracy was found to be affected by field size.

It was not possible to distinguish certain preprocessing alternatives as being superior to others and it was not statistically proved that a multitemporal classification produced higher accuracies than a monotemporal one.

Results did not indicate an evident influence of soil type on pixel values. It was concluded that classification results were not apparently influenced by variations in soil background.

The classification accuracy of some common Swedish crops was analysed. Generally, low accuracy values were obtained and it was concluded that a limited crop area estimation potential exists in southern Sweden. To fully ascertain classification possibilities, the study must be expanded in space and time and the classification accuracy of entire agricultural fields should be regarded. One very important matter to consider in subsequent studies is the development of techniques that discriminate edge pixels and pure pixels and classify them differently.

8

ESTIMATION OF CROP AREA BY MEANS OF LANDSAT MSS DATA

The classification accuracy of some common Swedish agricultural crops was calculated in section 7. The results were generally not encouraging and did therefore not indicate a promising crop area estimation possibility in the south of Sweden.
Results of section 6 indicated that it cannot be excluded that LANDSAT data have the capability of providing useful information for yield estimation purposes.

The objective of the present section was to:

(i) verify the conclusion drawn in section 7 by performing large crop area estimation studies

(ii) increase the knowledge of methods associated with estimation of scene attributes given the measured responses of the scene

8.1 BACKGROUND

8.1.1 Estimation of crop area

The Large Area Crop Inventory Experiment (LACIE) constituted an early effort to perform crop area estimations (MacDonald and Hall 1980). Subsequent studies concerning this matter are presented by e.g. Hixon et al. (1982), Chhikara (1984), Cook et al. (1984), Lennington et al. (1984), Ryeson et al. (1985), Moreira et al. (1986), McCloy et al. (1987). Moreira et al. (1986) estimated wheat areas in Brazil by means of aerial photographs and LANDSAT MSS data. Using a Gaussian maximum-likelihood classifier, the authors concluded that digital classification of single-date LANDSAT MSS data provided reasonably accurate results for wheat-area estimation (7.5 per cent overestimate) and that improvements in area estimation accuracy could be expected by using LANDSAT TM and SPOT data. McCloy et al. (1987) monitored rice areas in New South Wales. Regression analysis between estimated and observed rice area yielded coefficients of determination around 85-95%.

8.1.2 Estimation of quantitative scene attributes

The remote sensing problem is to estimate scene attributes, given the measured responses of the scene (Kimes 1983). In studies of vegetation and crops the two main attributes of interest are often kind of crop and amount. The classic quantitative estimation approach is to develop a deterministic model which relates the sensor response to underlying scene attributes. However, due to the presence of several unknown or unmeasurable variables and a high degree of experimental errors, the classical approach in many cases is not directly applicable. Therefore, reseachers have favoured simpler empirical models, usually regression-based calibration relationships between remotely sensed data and vegetation amount (Curran and Williamson 1987b).

The statistical techniques often used for relating remotely sensed data to vegetation variables are calculation of the correlation coefficient and regression. The regression technique is frequently used to fit an equation which can be inverted to supply transformation of remotely sensed data into expected vegetation variables. However, the "traditional" regression technique is unsuitable for this purpose in several cases and other techniques should be applied according to Curran and Hay (1986). The two reasons for this are related to (i) specification and (ii) assumptions concerning the nature of the data set (errors of measurements, Curran and Hay 1986).

When correlating two variables (X and Y) in a scatter plot, and the correlation coefficient between these does not equal 1, two lines may be fitted by regression methods - the regression of X on Y and the regression of Y on X . Thus, when $r_{xy} \neq 1$ then $Y = a + bX \neq X = \frac{Y - a}{b}$.

It is commonly argued that the equation should use the independent (causal) variable to predict the dependent (resultant) variable. As regards remote sensing applications, the independent variable is the Earth's surface and the dependent variable is the sensor response (Curran and Hay 1986).

Errors of measurement may be separated into errors in the measurement of ground variables, errors in the measurement of remotely sensed data and errors in the physical correlation of ground variables and remotely sensed data (Curran and Hay 1986). Errors in the measurement of ground variables are connected to the design of the measurement device and to the sampling procedure. Several error sources are related to the measurement of remote sensing data. These sources are reviewed by Curran and Hay (1986). The most important type of errors in the physical correlation of ground variables and remotely sensed data is misregistration in space and time.

In the literature the principal types of mistakes concerning inversion of regressions are reported as associated with:

(i) assumptions that X on Y = Y on X when $r_{xy} \neq 1$
(ii) assumptions that ground data have no error
(iii) assumptions that error in the remotely sensed

data is not related to the magnitude of X.

The first step towards remote sensing based empirical estimation of the Earth's features is usually to establish a model of the relationship between remote sensing data (Y) and an Earth variable (X, e.g. amount of biomass), i.e. X → Y. The second step includes inversion of this model to estimate X from Y, i.e. Y → X. Regression techniques proposed to solve the problem of correct inversion in remote sensing measurements include the inverse estimator and the Lwin-Maritz estimator (Pech et al. 1986), Wald's method, the "reduced major axis" and the "least-squares procedure" (Curran and Hay 1986).

Wald's method fits the regression line by dividing the observations into two halves on the basis of their x_i values (sample A and B). The line is fitted from

$$\beta = \frac{\bar{Y}B - \bar{Y}A}{\bar{X}B - \bar{X}A} \quad \text{and} \quad \alpha = \bar{Y} - \beta\bar{X}$$

The observations may also be divided into three parts using the highest and the lowest parts in the regression.

The logical base for this method is that the errors in X_i (assumed to sum to zero) will be compensated within the two sub-sets of data. The main doubts of the approach are (i) β and α will be associated with large sample variance and (ii) the method does not solve the specification problem.

The "reduced major axis" minimizes the sum cross products of the differences on both axes $(y_e-y_0)(x_e-x_o)$. Conventional regression minimizes the sum of the squares of the differences vertically (ye-y0) in the case of Y on X and horizontally (x_e-x_0) in the case of X on Y. Thus, the line may be fitted from

$$\bar{\beta} = \frac{\sigma y}{\sigma x} \quad \text{and} \quad \bar{\alpha} = \bar{y} - \frac{\sigma y}{\sigma x}\bar{x}$$

According to Curran and Hay (1986:236), this method is designed to "handle those problems in which errors of similar magnitude are present in the x's and the y's and in which the terms dependent variate and independent variate have no real meaning".

Curran and Hay (1986:238) conclude that " where no estimates of measurement variances are available Wald's method or the "reduced major axis" is to be preferred to regression". If knowledge concerning the relative magnitudes of the errors in the X and Y measurements exists the least-squared procedure is recommended.

The least-square procedure assumes that it is possible to separate the variance in X and Y into two parts; a true variance and an error variance. The method is discussed in detail by Curran and Hay (1986).

Prediction of biomass and yield or large-area estimations of yield were performed by Barnett and Thompson (1982, 1983), Steven et al. (1983), Kamat et al. (1985), Weiser et al. (1986). Barnett and Thompson (1982) investigated large-area relations between LANDSAT data and end-of-season wheat yield. A fairly close connection between estimated and measured yield was observed. Barnett and Thompson (1983) derived spectral data from LANDSAT MSS and NOAA-6 AVHRR and correlated with end-of-season-yield. A linear relationship was obtained between winter and spring wheat yield and spectral data. This was valid for examinations both on a small and large scale. Kamat et al. (1985) measured canopy spectral reflectance (hand-held radiometer) and physiological variables for wheat, chickpea and mustard during a complete crop cycle. Regression analysis between grain and biological yield and vegetation indices integrated over the crop cycle produced values for r between 0.59 and 0.77. The authors concluded that the obtained high correlations between spectral parameters and biological and grain yield clearly established the potentials of remote sensing for prediction of dry matter production as well as final grain yield. The correlation coefficient for regression between crop variables and spectral data was significantly improved by including weather parameters as temperature, rainfall, day length, and solar radiation. Predicted dry matter was found to agree closely to the observed dry matter. Weiser et al. (1986) established equations between spectral data and total above-ground dry green phytomass for grasslands. These equations were used to estimate green phytomass from spectral data the subsequent year. Estimates of phytomass were well correlated with measured phytomass values (R^2 = 0.78-0.81).

8.2 MATERIAL AND METHODS

8.2.1 Material

The initial intentions were to use the LANDSAT MSS as well as the LANDSAT TM data presented in section 6 in the present section. However, since the coverage of the TM data was unsuitable for large area crop estimations, the study was restricted to MSS data. Thus, remote sensing data comprised LANDSAT MSS data acquired on 83-05-19, 83-06-20, 83-07-06 of the province of Scania (section 2).

Crop data consisted of information concerning location and yield for a number of individual fields of winter wheat, winter rye, barley, winter rape seed and sugar beet (section 6.3.1 and Tab. 6.1.a). Reports published by the Statistical Bureau of Sweden offered lists of crop area (ha) and crop yield (kg/ha) for harvest regions in the province of Scania (Fig. 8.1). This information excludes agriculture fields belonging to farms with a total area of less than 2.0 ha.

Figure 8.1 The borders of the harvest regions in the province of Scania.

8.2.2 Methods

8.2.2.1 Crop area estimation

The method used for studying large crop area estimation accuracy included the following steps:

(i) standardization of at-satellite spectral reflectance (sections 3-4)
(ii) characterization of classes (training)
(iii) classification of multitemporal reflectance data
(iv) digitizing of harvest regions
(v) integration of digitized and classified data.

At the DIPIX ARIES III image processing system fields of winter wheat, winter rye, barley, winter rape seed and sugar beet (Tab. 6.1.a) were identified by means of field maps. These fields were used for training. No areas were masked out prior to classification.

Preprocessed multitemporal satellite data (standardized reflectance, MSS1-MSS4) were classified using the maximum likelihood algorithm (ML) at the DIPIX ARIES III system (section 7.2.2.2).

The borders of the harvest regions (Fig. 8.1) were digitized at a digitizing table. Through connection between the digitizing system and DIPIX ARIES III the digitized data were transformed to the image processing system and integrated with the classfied satellite scenes. To reduce the data space needed only 44 harvest regions were considered (the 44 most western regions).

For each crop type and cultivation region the number of classified pixels were calculated and transformed to hectares (1 pixel= 0.25 ha).

8.2.2.2 Estimation of quantitative vegetation variables

Crop data of winter wheat, barley and sugar beet (Tab. 6.1.a) were randomly divided into two halves. One half was used to calibrate yield and spectral data and the other half was used to estimate yield by means of inverted regression equations.

In the present study errors in the measurement of ground variables, errors in the measurement of remotely sensed variables and errors in the physical correlation of ground variables and remotely sensed variables were not established. Curran and Hay (1986) recommend that where no information concerning measurement variances is present Wald's method and the RMA should be used.

Differences between estimated yield values were compared for the calibration method of X on Y, Y on X, Wald and RMA.

Analyses were performed on spectral radiance and standardized at-satellite spectral reflectance of the ratio MSS4/MSS2 and included mono- and multitemporal data. Monotemporal data concern the acquisition of July 06. The mean spectral value of all three acquisitions (May 19, June 20 and July 06) was used as multitemporal data.

Statistical analysis included linear correlation analysis and analysis of variance. The Pearson product-moment correlation (confidence limits at the 95% level) and the F-test were used (Ryan et al. 1985).

8.3 RESULTS

8.3.1 Large crop area estimation

The actual and the classified area (ha) of winter wheat, winter rye, winter rape seed, sugar beet and barley for 44 cultivation regions are listed in Tab. 8.1. The total actual and classified area (ha) for all harvest regions together are listed below.

	Actual area (ha)	Class. area (ha)	Difference (ha)		
Winter wheat	51 389	124 651	73 262	142.5 %	overest.
Winter rye	14 301	77 062	62 761	438.9 %	overest.
Winter rape	33 567	30 436	3 131	9.3 %	underest.
Sugar beet	30 212	24 659	5 553	18.4 %	underest.
Barley	102 304	89 124	13 180	12.9 %	underest.

The regression of 44 cultivation regions yielded the following regression equations:

Winter wheat- ACT.AREA= 217 + 0.336xCLASS.AREA
R-sq.= 33.4%, S_e= 800.1
r= 0.578 (0.379-0.775)

Winter rye- ACT.AREA= 211 + 0.065xCLASS.AREA
R-sq.= 7.9%, S_e= 258.6
r= 0.282 (0.010-0.554)

Winter rape- ACT.AREA= 103 + 0.954xCLASS.AREA
R-sq.= 95.6, S_e= 148.5
r= 0.978 (0.965-0.991)

Sugar beet- ACT.AREA= 116 + 1.030xCLASS.AREA
R-sq.= 93.6, S_e= 167.4
r= 0.967 (0.948-0.986)

Barley- ACT.AREA= 795 + 0.756xCLASS.AREA
R-sq. 41.2%, S_e= 764.9
r= 0.642 (0.469-0.815)

where

ACT.AREA= actual area (ha)
CLASS.AREA= classified area (ha)
R-sq.= the coefficient of determination
S_e= the standard error of the residuals

Scatter plots of the relationship between actual and classified area (ha) are presented in Figs. 8.2.a-e.

Table 8.1
The actual and classified area of 44 harvest regions.
Reg no.= harvest region (Fig. 8.1)
A= actual area (ha)
C= classified area (ha)
D= difference between A and C (ha).

Reg no.	A.	C.	D.	A.	C.	D.	A.	C.	D.	A.	C.	D.	A.	C.	D
	Winter wheat			Rye			Rape seed			Sugar beet			Barley		
351	2	982	980	61	651	590	0	1	1	0	10	10	1031	1047	16
361	2	341	339	8	611	603	0	0	0	0	5	5	370	532	162
362	2	435	433	8	1560	1552	0	0	0	0	3	3	370	1076	706
363	32	801	769	66	2479	2413	0	2	2	9	28	18	676	2489	1813
364	243	2790	2547	389	2426	2037	79	56	-23	278	314	36	2584	3645	1061
365	0	1156	1156	7	544	537	0	3	3	1	3	2	882	1093	211
371	209	2794	2585	51	859	808	125	80	-44	16	8	-8	1589	2047	458
372	60	2305	2245	20	451	431	22	13	-9	12	12	0	1359	1411	52
373	112	2969	2857	56	656	600	132	77	-54	40	171	130	1719	1091	-628
374	0	1845	1845	24	464	440	0	4	4	3	6	3	1241	1450	209
381	306	3081	2775	322	1687	1365	58	75	17	165	101	-64	2604	3209	605
382	805	2969	2164	434	676	242	173	102	-70	245	90	-154	2092	2177	85
392	57	1798	1741	268	559	291	38	9	-28	189	74	-115	2248	2239	-9
401	2629	1986	-642	245	3551	3306	946	651	-294	781	1003	222	2560	2029	-531
402	1445	1333	-112	188	2977	2789	272	228	-44	686	534	-151	2415	2471	56
403	926	1076	150	150	1774	1624	93	104	11	393	195	-197	1882	1785	-96
405	1593	886	-707	202	4840	4638	202	52	-149	561	546	-14	2788	3379	591
421	2059	2394	335	489	2549	2060	1278	887	-339	1537	1385	-151	2485	1720	-764
422	703	2017	1314	868	1412	544	754	820	66	916	1161	245	1583	1324	-258
423	1368	973	-394	503	5030	4527	961	669	-291	1303	1416	113	2930	2292	-637
424	108	156	48	83	1936	1853	25	16	-9	155	108	-46	2432	1726	-705
431	3542	5858	2316	480	1603	1123	2490	2593	103	2638	2435	202	4140	2638	1501
432	2255	4716	2461	310	1537	1227	1872	1879	7	2340	2241	-99	3175	1693	1481
433	1413	1428	15	444	4019	3575	1028	492	-536	839	467	-371	1874	1447	-426
434	1516	2051	535	308	1292	894	1143	869	-273	1339	1175	-163	2083	2086	3
435	2487	4291	1804	730	1342	612	2112	2100	-11	2236	1855	-380	3237	1765	-1471
441	1835	2648	813	830	2500	1662	1300	1194	-105	817	577	-240	2029	988	-1041
442	2410	3821	1411	609	3400	2791	1129	1102	-26	568	196	-371	2514	2330	-183
443	1454	2974	1520	293	996	703	1071	1051	-19	1147	887	-259	2118	1150	-968
444	2064	5082	3018	455	1095	640	1259	1075	-184	870	341	-529	3214	1907	-1307
445	987	3283	2296	225	1578	1353	618	711	93	416	287	-128	1760	100	-758
451	324	3018	2694	1220	2273	1053	502	429	-72	190	233	43	1985	2667	682
452	771	5396	4625	414	1503	1089	487	335	-151	420	76	-343	3829	3323	-506
454	778	4538	3760	156	520	364	669	426	-242	311	106	-205	3339	2185	-1154
455	616	4455	3839	145	1048	903	794	652	-141	487	336	-150	3582	2201	-1380
461	1509	1919	410	354	1061	707	913	650	-262	811	498	-313	2243	2487	244
462	810	3329	2519	681	1271	590	968	1099	131	712	323	-388	2189	1893	-295
463	2116	4487	2371	160	565	405	1662	1629	-32	1538	1164	-373	2863	1712	-1150
464	2166	6645	4479	504	1765	1261	1584	1428	-154	672	464	-207	4644	2513	-2130
465	631	2303	1677	619	1944	1325	702	482	-220	402	223	-178	1752	1921	169
471	1511	1881	370	315	1115	800	1118	1033	84	891	856	-34	1828	1295	-532
472	3072	4940	1868	241	3499	3258	2458	2650	192	1484	1295	-188	3526	3203	-323
473	2338	6255	3917	259	2336	2077	1102	985	-116	599	634	35	4312	4822	510
474	2123	4223	2100	99	1091	992	1428	1707	279	1045	954	-90	2228	1650	-577

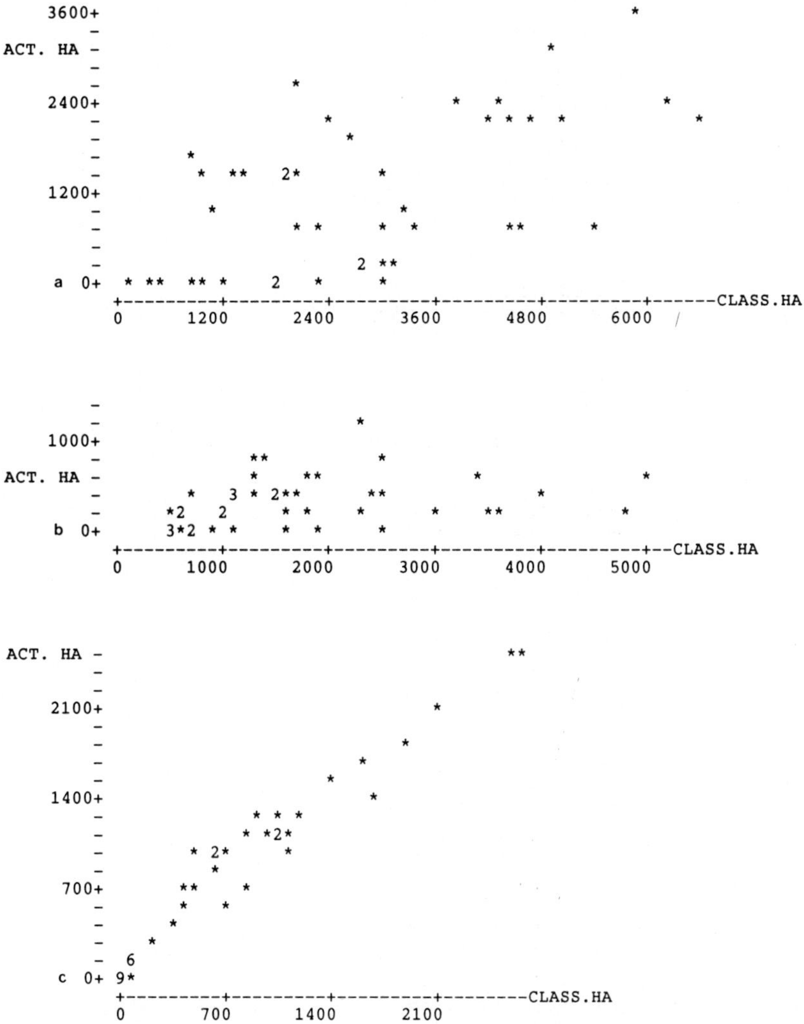

<u>Figure 8.2</u> The relationship between actual (ACT) and classified (CLASS) crop area. HA= hectares
(a)= winter wheat, r= 0.578, n=44
(b)= winter rye, r=0.282, n=44
(c)= rape seed, r=0.978, n=44

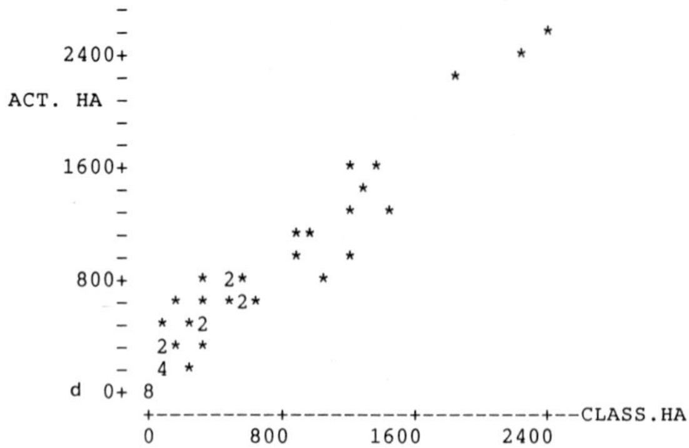

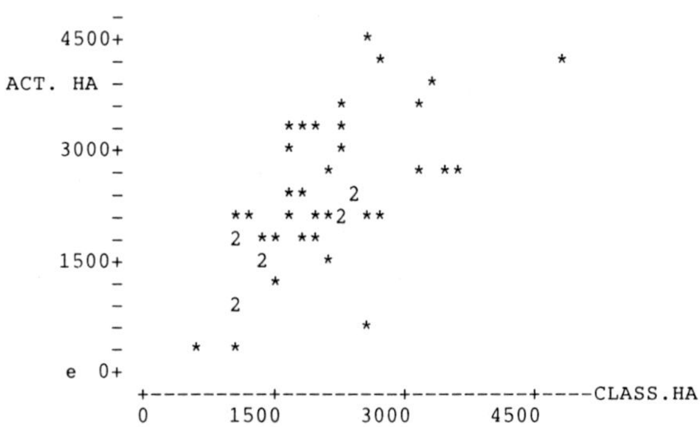

Figure 8.2 cont.
d= sugar beet, r=0.967, n=44
e= barley, r= 0.642, n=44

Results show that winter wheat and winter rye are associated with very high total overestimations (142.5% and 445.2%) as well as low correlation coefficients of the relationship between actual and classified area. The area estimation study of barley produced somewhat stronger agreememt between actual and classified area compared with rye and wheat. The estimation accuracy for sugar beet and rape seed was rather good with correlation coefficients around 0.97 and total underestimations of 9.3% and 18.4%.

Winter rye was clearly separated from other cereals in section 6 and in section 7 this crop produced fairly high classification accuracies. In the present section a large overestimation was obtained. Presumably, this overestimation is caused by classification of non-agriculture land. Application of non-agriculture land cover masks prior to classification might have increased the area estimation accuracy of rye. Results of sections 6 and 7 indicate that winter wheat and barley are often spectrally confused and associated with rather low classification accuracies. Spectral data of rape seed and sugar beets differed significantly from each other as well as from other crops on several acquisition dates (section 6). The yellow colour of the rape seed flowers and the unique phenological curve of sugar beets probably have an influence on the significant difference between these crops as well as on the rather good area estimation accuracy.

Thus, based on a "traditional" multitemporal ML-classifier of MSS data the following conclusions were made for the investigated provinces in Sweden:

- for cereals like winter wheat, rye and barley a low area estimation possibility exists
- the area estimation potential of rape seed and sugar beet is high
- further studies of crop area estimation accuracies should include a non-agriculture mask.

8.3.2 Estimation of quantitative vegetation variables

Fig. 8.3 shows the application of four calibration methods to the winter wheat data of Tab. 6.1.a. Data used for this consist of half the data set presented in Tab. 6.1.a. The figure illustrates the differences of estimated yield values when applying regressions of X on Y, Y on X, Wald's method and the reduced major axis. The largest difference was between X on Y and X on Y while Wald's method and RMA produced yield values between these extremes. Practical application of Wald's method and RMA of half the data set presented in Tab. 6.1.a produced the following fitted forms:

winter wheat,	Wald:	$Y= 1.57 + 0.022X$	(8.1)
	RMA:	$Y= 1.41 + 0.024X$	(8.2)
barley,	Wald:	$Y= 1.58 + 0.028X$	(8.3)
	RMA:	$Y= 0.34 + 0.063X$	(8.4)
sugar beet,	Wald:	$Y= 0.73 + 0.011X$	(8.5)
	RMA:	$Y= 0.67 + 0.011X$	(8.6)

Forms used for estimation were thus

winter wheat,	Wald	$X= -71.4 + 45.5Y$	(8.7)
	RMA	$X= -58.7 + 41.6Y$	(8.8)
barley,	Wald	$X= -56.4 + 35.7Y$	(8.9)
	RMA	$X= -5.39 + 15.9Y$	(8.10)
sugar beet,	Wald	$X= -66.3 + 90.9Y$	(8.11)
	RMA	$X= -60.9 + 90.9Y$	(8.12)

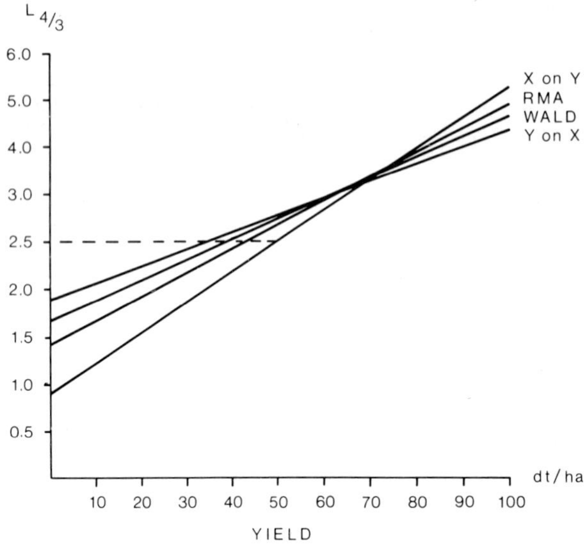

Figure 8.3 Application of four calibration methods.

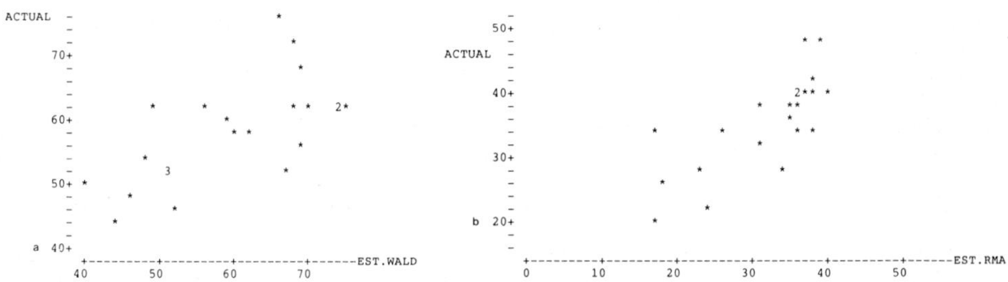

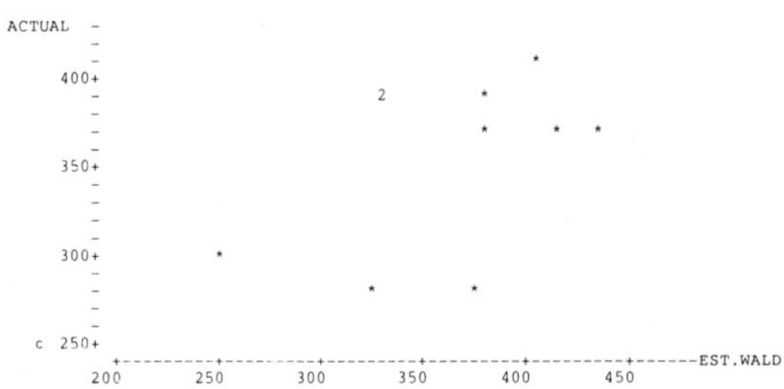

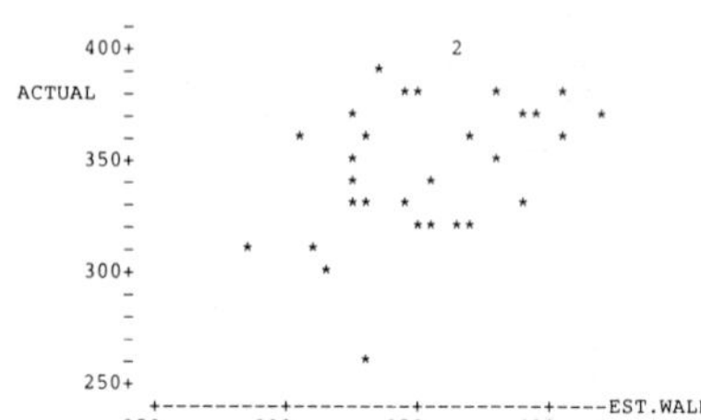

Figure 8.4 The relationship between actual and estimated yield. (a)= winter wheat, (b)= barley, (c)= sugar beet.

Figure 8.5 Regional relationship between actual and estimated sugar beet yield.

where
X= at-harvest yield (10^2kg/ha)
Y= MSS4/MSS2 (unitless at-satellite spectral reflectance).

The at-satellite spectral reflectance (MSS4/MSS2) of the other half of the data set was integrated in equations 8.7-12 and estimated yield values were obtained.

The relationship between actual and estimated yield is plotted in Figs. 8.4.a-c. The correlation coefficients between actual and estimated yield for Wald's method and RMA were:

winter wheat	0.671
barley	0.755
sugar beet	0.430

The mean at-satellite spectral reflectance (MSS4/MSS2) of all pixels classified as sugar beet within each harvest region was used in equations 8.11-12 and an estimated yield value for each region was obtained. The correlation coefficient between estimated and actual yield was 0.414 (0.124-0.704). A scatter plot of the relationship is presented in Fig. 8.5. Only for 30 of the 44 harvest regions yield data were available.

The above mentioned equations, correlation coefficients and scatter plots are valid for monotemporal at-satellite spectral reflectance (July 06). No significant difference was obtained when applying spectral radiance or multitemporal data.

8.4 STUDY REMARKS

* Radiometric errors may have arisen due to the geometric correction.

* The classification accuracy was not calculated in this section.

* The problem of mixed pixels was not regarded (Chhikara 1984, Lennigton et al. 1984) and the applied classifier is probably not the most optimal one. A contextual classifier might have led to different results and conclusions.

* Training was performed on a weak statistical material (too few fields, specially for the sugar beet data).

* The spectral data used in Fig. 8.3 is the ratio MSS4/MSS2 of spectral radiances. As the solar spectral irradiance differs between bands, a ratio between exoatmospheric reflectances is the most adequate one.

8.5 SUMMARY

The objective of the studies presented in this section was to:

(i) analyse crop area estimation accuracies

(ii) examine methods associated with estimation of scene attributes given the measured responses of the scene.

Winter wheat and rye were strongly overestimated while the estimated area of rape seed and sugar beet was very well connected with the actual area. It was concluded that for cereals like winter wheat, rye and barley a low area estimation possibility exists, while the area estimation potential of rape seed and sugar beet is high in the studied provinces of Sweden.

Various calibration techniques produced different values of estimated yield. The importance of applying correct regression methods was stressed.

9

ESTIMATION OF SUGAR BEET AREA BY MEANS OF LANDSAT TM DATA

One of the most important crops in the south of Sweden is sugar beet. The Swedish Sugar Company has an overall responsibility for cultivation and has great interest in finding methods for facilitating area estimation, yield prediction and monitoring of diseases.

This section (9) and the following one (10) present studies, ordered by the Company, of the possibility to monitor various sugar beet variables by means of remote sensing.

The aim of the work presented in this section was to analyse:

1. the area estimation accuracy on a large area scale as well as on an individual field scale

2. the relationship between actual and spectrally estimated sugar beet yield.

9.1 INTRODUCTION

9.1.1 Estimation of sugar beet yield by remote sensing

The rate of the increase of biomass per unit ground area i.e. the plant productivity has been related to the amount of solar energy intercepted by the plant (Gallagher and Biscoe 1978). The amount of light that may be intercepted by the plant is dependent on the amount of photosynthetically active biomass (GLAI) and its duration (De Wulf and Goossens 1986). Several authors have obtained a close relationship between spectral data and GLAI of plants (Badhwar et al. 1986, De Wulf and Goossens 1986, Hinzman et al. 1986).

The use of spectral GLAI measurements to estimate further rate of sugar beet plant growth, productivity and final yield is, however, associated with several problems. One is the complex relationships between meteorological parameters (too little rain, too much rain, cool days, frosty nights etc.) during various phenological stages and leaf development, root weight and sugar content in the root. Especially the root sugar content is dependent on meteorological factors (De Wulf and Goossens 1986). Spectral data for yield estimation was discussed in section 6.1.3 .

9.1.2 Review of literature

Remote sensing estimation of sugar beet biomass and productivity and studies of sugar beet diseases have been performed by for instance Steven et al. (1983), Andrieu (1984), De Wulf and Goossens (1986).

Steven et al. (1983) used a two-band spectrometer to measure solar radiation reflected from sugar beets. Measurement of light interception was performed by placing solarimeters at the base of and above the crop leaf cover. Leaf cover was determined from photographs over the crop. The authors found the light interception to be proportional to leaf cover and that the relationship was almost 1:1. The ratio IR/R increased with the percentage of ground covered by healthy green leaves and was used to determine the fraction of incident solar energy intercepted by the leaves. The measurements were used to determine future rates of growth and predict yield to within 6 per cent.

The possibility of using remote sensing data to detect rhizome in sugar beets was evaluated by Andrieu (1984).

De Wulf and Goossens (1986) used multitemporal spectral reflectance data based on colour infrared photography to measure sugar beet variables (GLAI, cover percentage, fresh root weight and sugar content). GLAI was determined by placing a transparant dot grid on top of the photographic paper. Statistical relationships between spectral data and agronomic parameters were derived for one growing season and used for prediction in subsequent growing season. The authors found that GLAI and cover percentage could be estimated rather precisely but that further experiments using additional data were needed to confirm these results. Due to meteorological factors, results did not manifest simple relationships between crop canopy and root biomass. It was concluded that, likely, estimated GLAI may be used in yield prediction models and that the solution is a procedure that incorporates both meteorological and remote sensing parameters.

9.2 MATERIAL AND METHODS

9.2.1 Material

The sugar beet cultivation in Sweden is controlled by the Swedish Sugar Company. The individual farmers cultivating sugar beets thus have a contract with the company concerning the area (ha) of sugar beets that may be sown. By field measurements of randomly selected farms the Company each year checks the correspondence between the contracted area and the area that has been sown. For the sugar-refinery of Jordberga (Fig. 9.1.a) approximately 350 farms were controlled in 1986. These farms had a total contracted area of 2489.2 ha while the "real" area through field checks was found to be 2503.09ha. The "overcultivation" of these farms was thus around 0.6%. The sugar beet data used in the present study were obtained from the Company and concerned the sugar-refinery of Jordberga (Fig.9.1.a).

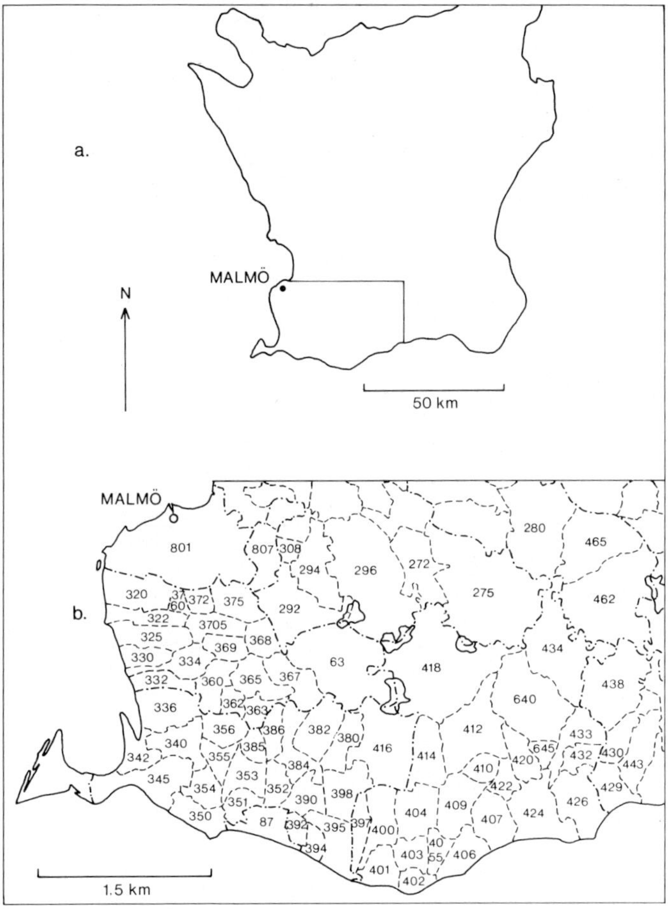

Figure 9.1 a. The area of the sugar-refinery of Jordberga. b. The cultivation regions of the sugar-refinery of Jordberga.

The total area of agricultural land cover for this region is 62 475ha. The following information valid for the harvest of 1986 was received:

(i) field measured area (ha) of 77 individual sugar beet fields (randomly selected)

(ii) contracted sugar beet area (ha) of 76 cultivation

regions and a map of the borders of these regions (Fig. 9.1.b)

(iii) at-harvest sugar beet yield of 77 individual fields.

Remote sensing data included a LANDSAT 5 TM acquisition (Q4) from 1986-09-24 (section 2). Preprocessing was performed on a DIPIX ARIES III system and comprised (i) calculation of spectral radiance, (ii) calculation of at-satellite spectral reflectance, and (iii) geometric correction.

The south-west part of the scene was rather cloudy. However, due to lack of reference data, no correction and adjustment was performed. The following regions (Fig. 9.1.b) were visually associated with clouds: 253, 345, 350, 351, 352, 354, 355.

9.2.2 Methods

9.2.2.1 Area estimation

Initially, information of the location of 32 randomly selected sugar beet fields was obtained from the Sugar Company. By means of field maps the fields were identified on the monitor of the DIPIX ARIES III image processing system. These fields were used in training to characterize the sugar beet class. Tab. 9.3 presents the size of the fields used for training. Using the maximum likelihood (M-L) algorithm (section 7.2.2.2) on the DIPIX ARIES III system pixels were classified as sugar beet or left unclassified. A post classification "clean-up" filter was applied. Classification included TM 1-5 and TM7 and was performed on spectral radiance and at-satellite spectral reflectance. The whole scene was included in the classification and thus no surfaces were masked out.

The position and measured area (ha) of further 45 randomly selected fields were obtained from the Company. The fields were located at the DIPIX monitor and for each field the number of pixels classified as sugar beet was calculated and transformed to hectares (1 pixel= 0.04 ha).

Through connection between the DIPIX ARIES III system and a digitizing table the borders of the 76 cultivation regions (Fig. 9.1.b) were digitized and integrated with the classified image. Appendix H illustrates a printout of this. For each cultivation region the number of pixels classified as sugar beet was calculated and transformed to hectares. The classified area (ha) was compared with the contracted area (ha). Information concerning the "true" area sown with sugar beet within each region is not available.

9.2.2.2 Yield estimation

The direct relationship between yield and spectral data as well as the accuracy of estimation models based on only spectral data were examined. This was performed by the following steps:

1 - Modelation of the relationship between satellite data and sugar beet yield.

This was carried out by measuring the spectral radiance and the at-satellite spectral reflectance of the 32 fields used for training and calculating the Pearson product moment correlation coefficient with 95% confidence intervals between the yield of these fields and the satellite data. Single bands as well as simple ratios (TM4/TM1, TM4/TM2, TM4/TM3, TM4/TM7, TM5/TM3, TM4/TM5) were used.

2 - Inversion of "step 1" to estimate yield from spectral data.

The "Wald and Bartlett" method and the "reduced major axis" approach (Curran and Hay 1986, section 8) were applied to estimate yield from spectral data.

3 - Assessment of estimation result.

The spectral radiance and reflectance of further 45 fields were measured. These data were applied to the inverted models of "step 2" and estimated values of the yield were obtained. For these 45 fields the estimated yield was compared with the true yield received from the Sugar Company.

9.3 RESULTS

9.3.1 Large-scale area estimation.

The total contracted area of sugar beets was 9890 ha. The Company considers that this agrees to a harvested area of 9941 ha.

Tab. 9.1 presents the contracted and classified area of sugar beets for each cultivation region (Fig. 9.1.b). The total classified area was 11 491ha and thus the difference between classified and contracted area was 1601ha while the difference between classified and "harvested" area was 1550ha. This corresponds to an "overestimation" of 16.2% and 15.6%.

The mean difference between classified and contracted area was approximately 50 ha. 25 of the 77 cultivation regions were "underestimated" with a total area of 41.8 ha. The total "overestimated" area for the other 52 regions was 59.3 ha.

Table 9.1 The contracted and classified area (ha) of sugar beet within 76 cultivation regions.

No.	Co.	Cl.	D.	No.	Co.	Cl.	D.
63	144.5	193.4	48.9	390	128.2	148.7	20.6
87	25.1	18.4	-6.7	392	30.6	26.4	-4.3
272	20.5	144.6	124.1	394	36.2	32.5	-3.7
275	14.7	53.6	38.8	395	211.8	191.7	-20.1
280	23.7	145.8	122.1	397	180.4	182.9	2.5
292	171.7	292.3	120.6	398	275.6	254.9	-20.6
296	72.6	117.3	44.8	400	252.0	309.4	57.4
308	6.4	37.0	30.7	401	148.9	158.7	9.8
320	142.9	139.6	-3.3	402	25.1	37.6	12.5
322	109.3	134.3	24.9	403	205.8	156.8	-49.0
325	201.5	276.6	75.0	404	321.7	344.3	22.6
330	85.6	126.7	41.1	406	136.0	111.2	-24.8
332	95.2	115.6	20.4	407	211.2	111.6	-99.6
334	167.8	177.8	10.0	409	309.7	326.4	16.7
336	247.7	222.2	-25.5	410	112.4	126.5	14.1
340	123.0	84.0	-39.0	412	190.6	273.3	82.7
342	57.3	74.4	17.1	414	165.2	269.3	104.0
345	152.5	37.4	-115.0	416	336.2	377.6	41.4
350	138.3	73.0	-65.3	418	11.3	106.6	95.3
351	107.9	47.2	-60.7	420	150.3	214.7	64.5
352	108.9	69.9	-38.9	422	75.9	91.6	15.7
353	193.0	92.9	-100.0	424	299.4	227.6	-71.8
354	108.9	9.0	-99.9	426	268.3	360.6	92.3
355	147.8	101.4	-46.8	429	183.4	205.0	21.6
356	182.3	180.1	-2.1	430	115.0	193.6	78.6
360	131.6	166.7	35.1	432	69.8	161.9	92.1
362	75.5	70.6	-4.9	433	95.9	142.1	46.2
363	60.4	91.7	31.3	434	145.0	287.6	142.6
365	74.4	154.4	80.0	438	143.2	140.8	-2.3
367	73.8	149.8	76.0	443	109.5	189.4	79.9
368	35.5	131.2	95.7	462	136.9	112.5	-24.3
369	103.0	127.0	24.0	465	10.5	61.7	51.2
372	49.0	64.5	15.5	640	260.9	334.8	73.9
375	48.3	114.7	66.4	645	93.3	69.9	-23.3
380	93.2	125.6	32.4	801	93.3	95.3	2.0
382	98.5	179.8	81.3	807	53.8	158.9	105.1
384	335.5	241.0	-94.5	3705	93.5	95.6	2.1
386	99.3	113.5	14.2	4055	76.5	105.0	28.5

No.= cultivation region (Fig. 9.1.b)
Co.= contracted area (ha)
Cl.= classified area (ha)
D. = difference between Co. and Cl. (ha)

A least-fit regression for the 77 regions yielded a 59.2 coefficient of determination, a regression equation of

CONTR.AREA(ha)= 19.7 + 0.731xCLASS.AREA(ha)

S_e= 53.1

where
contr.area(ha)= contracted area
class.area(ha)= classified area

A scatter plot of the relationship is shown in Fig. 9.2.a.

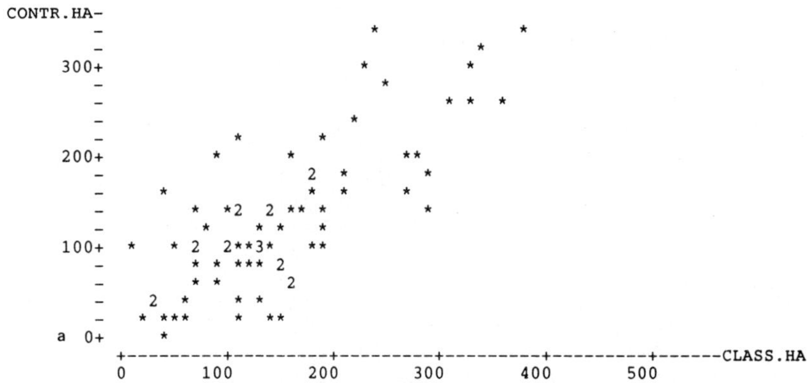

Figure 9.2.a The relationship between contracted and classified sugar beet area (ha). r= 0.769, n= 76. Contr.ha= contracted area, class.ha= classified area.

Of the 25 regions that were "underestimated" 7 were associated with visible clouds. When these 7 regions were excluded from the analysis, the total contracted area was 8933 ha and the "harvested" area 8984 ha while the total classified area was 11061 ha. Thus, a difference of 2128 ha and 2080 ha corresponding to a total "overestimation" of 23.82% and 23.15% was obtained.

The least-fit regression for this case (70 regions) produced with a 68.5 coefficient of determination, a regression of

CONTR.AREA(ha)= -4.0 + 0.833xCLASS.AREA(ha)

S_e= 48.7

The scatter plot for this case is presented in Fig. 9.2.b.

Thus, when regions with clouds were excluded, the total

"overestimation" increased with approximately 7.5% while the correlation coefficient between contracted and classified area increased from 0.769 (0.677-0.860) to 0.828 (0.754-0.902).

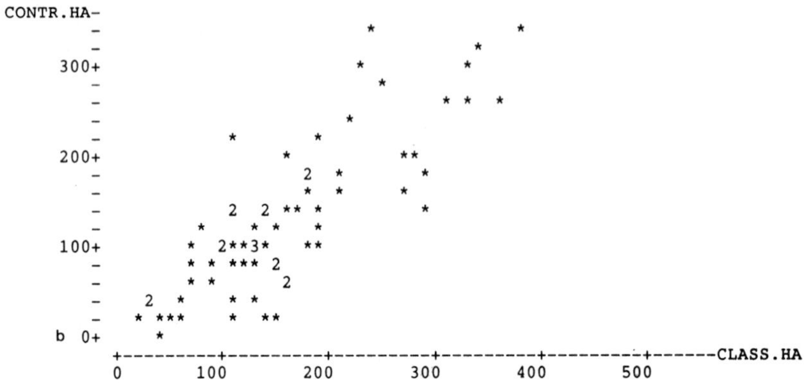

Figure 9.2.b The relationship between contracted and classified sugar beet area (ha). r=0.828, n= 69. Contr.ha= contracted area, class.ha= classified area.

The above results concern classification based on at-satellite spectral reflectance. No significant difference was obtained between classification based on spectral radicance and at-satellite spectral reflectance.

Most crops in Scania are harvested in August and sugar beet is the only crop that is harvested in late autumn. The per-pixel classification was performed on all pixels without masking any surfaces. It cannot be excluded that increased area estimation accuracies would be obtained if e.g. non-agricultural areas were masked out prior to classification. In an actual operational use of satellite data for large area sugar beet estimation this is recommended.

The present study as well as the area estimation study of section 8 indicate a rather strong connection between actual and estimated sugar beet area. Encouraging results were obtained for TM as well as for MSS data and for various acquisition dates (July and September). Probably satellite data may be integrated advantageously in the activity of the Sugar Company.

9.3.2 Field-scale area estimation

Tab. 9.2 exhibits the area (ha) of 45 randomly selected individual fields and the classified area (ha) of these fields. The mean area of the fields was 3.62 ha and the classified mean area was 3.98 ha. Of the 45 fields, 4 were smaller than 1.0 ha. Fields smaller than 1.0 were not classified as sugar beet, and as can be seen in Tab. 9.2 the classified area of these fields was 0 ha.

Table 9.2 The actual and classified area (ha) of 45 sugar beet fields

A.	C.	D.	A.	C.	D.
0.88	0.0	-0.88	8.30	8.76	0.46
6.57	11.04	4.47	1.14	2.64	1.50
11.32	10.80	0.52	6.30	5.96	-0.34
3.78	4.32	0.54	2.10	1.08	-1.02
2.64	3.40	0.76	1.14	2.84	1.70
2.73	3.24	0.51	3.77	3.76	-0.01
2.28	3.00	0.72	3.93	6.08	2.15
1.69	0.92	-0.77	1.07	0.64	-0.43
2.71	3.16	0.45	1.32	0.68	-0.64
3.07	3.68	0.61	2.28	2.10	-0.18
8.01	7.20	-0.81	2.12	0.28	-1.84
5.63	7.00	1.37	0.33	0.00	-0.33
2.33	1.12	-1.21	5.31	9.80	4.49
0.95	0.00	-0.95	5.96	6.04	0.08
4.11	5.40	1.29	1.98	1.12	-0.86
0.64	0.00	-0.64	7.42	8.76	1.34
3.17	4.56	1.39	5.36	3.36	-2.00
3.99	6.88	2.89	4.50	5.08	0.58
2.05	0.24	-1.81	1.68	2.20	0.52
3.21	4.08	0.87	2.97	1.28	-1.69
7.00	7.08	0.08	2.75	3.84	1.09
3.00	4.28	1.28	3.53	4.76	1.23
5.79	6.72	0.93			

A.= actual areal (ha)
C.= classified area (ha)
D.= difference between actual and classified area

Scatter plots of the relationship between actual area (ha) and classified area (ha) are presented in Figs. 9.3.a-b. All the 45 fields are included in Fig. 9.3.a (r=0.888, 0.826-0.949) while the 4 zero-classified fields are excluded in Fig. 9.3.b (r=0.868, 0.792-0.943).

The mean percentage of misestimation was 45. As can be observed in Fig. 9.3.c the percentage of misestimation decreased with increased field size.

This "automatic" field area estimation was compared with a "manual" field area estimation. The area of 32 fields was measured by marking the limits of the fields at the DIPIX

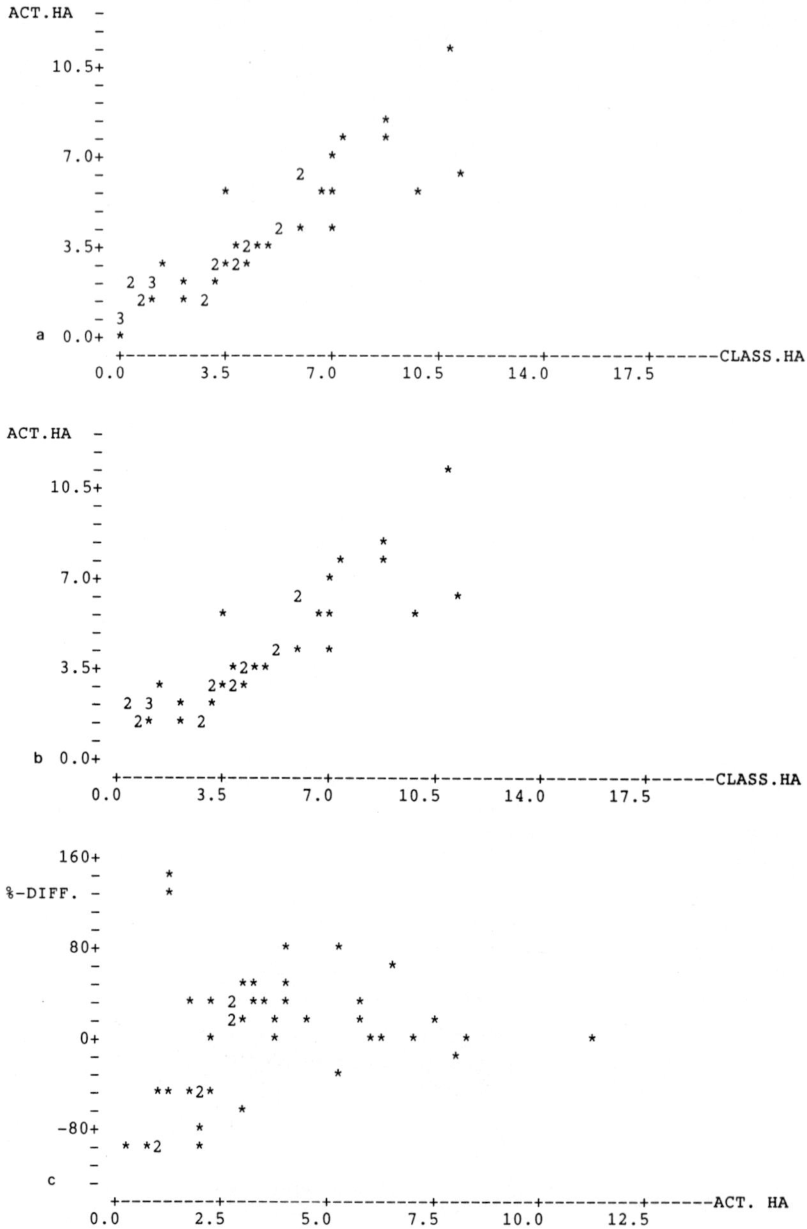

Figure 9.3 The relationship between actual (ACT) and classified (CLASS) area. HA= hectares.
(a), r= 0.888, n=45
(b), r= 0.868, n=41
(c), the percentage of misestimation

AREAS III monitor. This method is thus dependent on the operator's personal decision of the field border location. The actual area (ha) for the 32 fields and the "manually" measured area (ha) for these fields are tabulated in Tab. 9.3. Fig. 9.4.a displays a scatter plot of the relationship (r= 0.992, 0.986-0.998).

The mean difference between actual and manually estimated field size was approximately 6% (Fig. 9.4.b).

Table 9.3 The actual and "manually" measured area of 32 sugar beet fields.

A.	M.	D.	A.	M.	D.
14.11	14.67	0.52	14.42	14.42	0.00
19.22	19.95	0.73	4.26	4.32	0.06
6.57	6.77	0.20	12.37	15.20	2.83
7.85	8.10	0.26	7.47	6.88	-0.59
10.06	10.41	0.35	4.25	5.04	0.79
10.41	10.77	0.36	3.86	4.32	0.46
16.01	16.61	0.59	3.93	3.88	-0.05
6.71	6.92	0.21	10.17	11.12	0.95
9.74	10.08	0.34	5.35	5.48	0.13
6.70	7.72	1.02	14.80	15.28	0.48
5.55	5.44	-0.11	1.42	1.24	-0.18
4.20	3.88	-0.32	8.26	8.40	0.14
9.41	9.04	-0.37	15.44	14.80	-0.64
7.35	7.34	-0.01	2.59	2.92	0.33
14.17	15.12	0.95	8.35	8.16	-0.19
10.18	10.64	0.46	4.72	4.40	-0.32

A.= actual areal (ha)
M.= measured area (ha)
D.= difference between actual and measured area (ha).

The mean size of the fields included in the "automatic" field area estimation was 3.62 ha, while the mean size of the fields included in the "manual" field area estimation was 8.74 ha. Since the area estimation accuracy is dependent on the field size it is not possible to compare the accuracy of the "automatic" and the "manual" area estimation method.

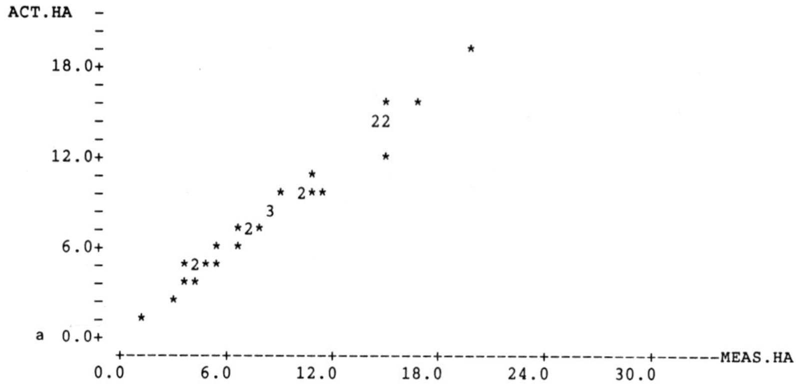

Figure 9.4.a The relationship betwen actual and "manually" measured field area (ha). r= 0.992, n=32. Act.ha= actual area, meas.ha= manually measured area.

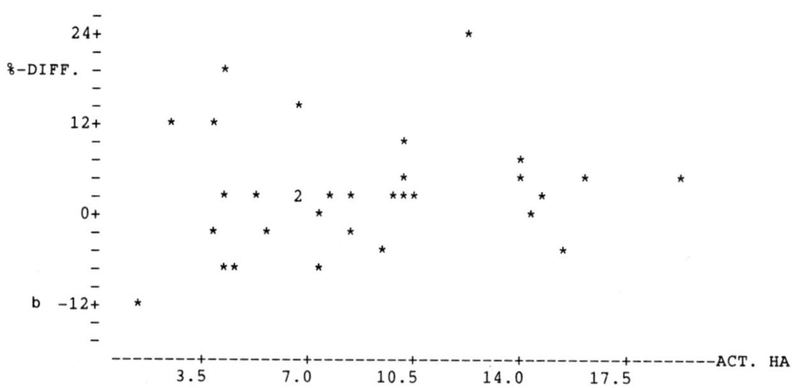

Figure 9.4.b The difference (%) between manually measured area and actual area (ha). Act.ha= actual area.

The "manual" as well as the "automatic" field area estimation method produced a good relationship between observed and estimated field area (correlation coefficients around 0.85-0.99). As regards field area estimation, it is concluded that satellite data may be a useful instrument. Thus, it may be profitable to use satellite data in the field area measurements performed by the Sugar Company each year. A possible approach is to integrate (i) a sugar beet classification, (ii) a manual sugar beet area estimation performed at an image processing system and (iii) traditional field measurements of very few fields. The economic condition of satellite-based field area estimation should be evaluated against the method used today.

9.3.3 Yield estimation

For 32 randomly selected fields the relationship between at-harvest yield and satellite data calculated as spectral radiance and at-satellite spectral reflectance for TM1-5 and TM7 and simple ratios was analysed. The highest correlation coefficient (0.359, 0.057-0.661) was obtained for TM4/TM3 (at-satellite spectral reflectance). Yield is plotted against spectral data in Fig. 9.5. Analysis of variance between spectral data of fields associated with yield values between 30-45 10^3kg/ha and 46-60 10^3kg/ha did not produce statistically significant differences.

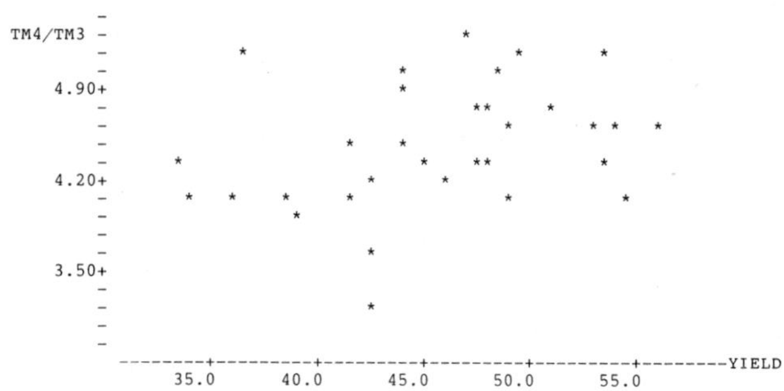

Figure 9.5 The relationship between yield and TM4/TM3 (at-satellite reflectance). r= 0.359, n=32.

When using the "Wald and Bartlett" method (Wald) and the reduced major axis approach (RMA) fitted forms were created;

Wald: $Y= 3.9 + 0.06X$ (9.1)
RMA: $Y= 1.4 + 0.12X$ (9.2)

Forms used for estimation were:

Wald: $X= -65.0 + 16.7Y$ (9.3)
RMA: $X= -11.7 + 8.3\ Y$ (9.4)

where
X= sugar beet yield (10^3kg/ha)
Y= TM4/TM3 (unitless effective at-satellite reflectance).

The at-satellite spectral reflectance (TM4/TM3) of 44 additional fields were used in equations 9.3-4 and estimated yield values for these fields were thereby obtained. Relationships between estimated and actual yield were associated with very low correlation coefficients (around r= 0.15). The yield data were divided into five classes (0-35,

36-40, 41-45, 46-50, 51-60 10^3kg/ha). The number of fields within each class correctly "classified" was measured. Very low values of per cent correctly classified were obtained.

Thus, results indicated a low connection between sugar beet yield and spectral data and between estimated and actual yield. The strong correlations between yield and spectral data found in section 6 could not be repeated in the present study. Section 6 included mid-season data (June-July) while the present study was performed on end-of-season data (end of September). The obtained differences may possibly be associated with the photosynthetic potential and the plant cover. Further studies of sugar beet yield and satellite data should concern mid-season acquisitions.

9.4 STUDY REMARKS

* No atmospheric correction was performed.

* Radiometric errors may have arisen due to the geometric correction.

* The problem of "mixed pixels" (Chhikara 1984) was not considered. A contextual classifier might have led to different results and conclusions.

* The possible influence of soil type variations was not considered.

9.5 SUMMARY

The aim of this section was to analyse:

1. the sugar beet area estimation accuracy on a large area as well as on an individual field scale

2. the relationship between sugar beet yield and satellite data as well as the relationship between "actual" and "estimated" yield.

The large area estimation study produced correlation coefficients between contracted and estimated area around r=0.8. The total "overestimation" was approximately 20%. It was concluded that satellite data may be useful concerning large area sugar beet estimation.

The mean area of 45 randomly selected individual sugar beet fields was 3.62 ha, while the mean classified area for these fields was 3.98 ha. The correlation coefficients between actual and classified field size were around r=0.85. A "manual" estimation of field sizes yielded correlation coefficients around r=0.99. Presumably, it can be favourable to integrate satellite data in the field area controls performed by the Company.

The relationship between late season LANDSAT data and sugar beet yield was weak. It was concluded that further studies of sugar beet yield and remote sensing data should concern mid-season acquisitions.

10

STUDIES OF THE RELATIONSHIP BETWEEN SUGAR BEET VARIABLES AND DIGITIZED COLOUR INFRARED AERIAL PHOTOGRAPHY

10.1 INTRODUCTION

In this section found statistical relationships between colour infrared aerial photography and various sugar beet variables are reported.

The work was performed together with Lu Jiaju who carried out the measurement of digital numbers. The study was published by Hall-Könyves and Jiaju (1987).

10.2 MATERIAL AND METHODS

10.2.1 Sugar beet data

The study was accomplished during the vegetation period of 1986 in a field belonging to The Swedish Sugar Company and located about 10km outside the city of Lund (Fig. 10.1.a) in Scania in the south of Sweden.

The field (approximately 150x500m) was composed of 13 sites grown with sugar beets (Fig. 10.1.b). Each site comprised a number of small plots (approximately 1-2m^2, Fig. 10.1.c). The different sites were subjected to various forms of growing conditions and cultivation and treatment systems. Within each site the growing condition and cultivation system varied among the plots which formed plot groups with similar subtreatment (Fig. 10.1.c).

Tab. 10.1.a gives information about the main type of growing condition for each site and the number of plots and plot groups within each site. For further information about the sugar beet sites see "Plansamling för Samarbetskommittens försök, 1986, Sockerbolaget, Jordbruksteknik".

The sowing was performed in April to May and the plots within each site were planted on the same day (except site 2). Two rows within each plot were harvested during normal harvest time (October/November). The yield of these two rows is according to the Sugar Company very well correlated with the total yield of the plot and thus constitutes a good measure of the total yield of each plot. Sugar mass (t/ha) was calculated from fresh root weight data (t/ha) and sugar content (%) analysis. The distribution of the yield (sugar mass t/ha) within each site is shown in Tab. 10.1.b.

The plant numbers of sites 4,5,7 and 12 was calculated on the same day as the aerial photograph was taken.

Table 10.1.b The yield distribution within each site.

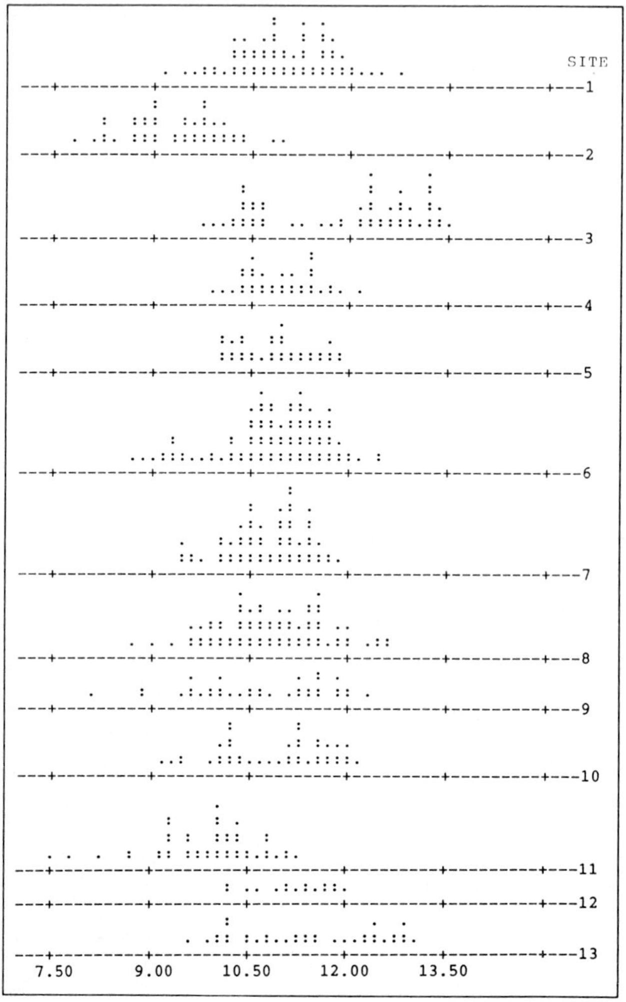

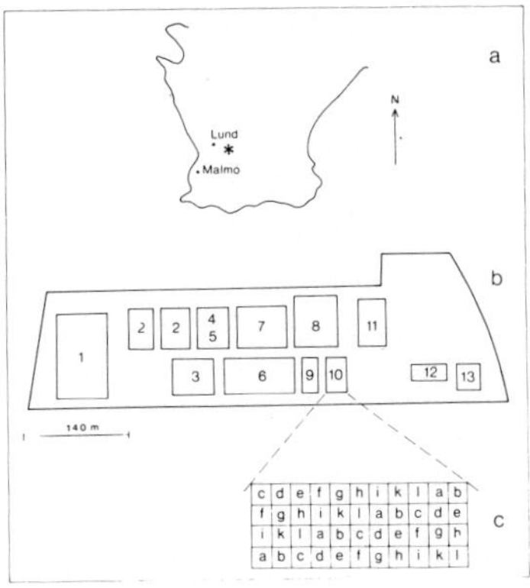

Figure 10.1. (a) Scania in the south of Sweden, *= the location of the field. (b) The location of the 13 sites within the field. (c) Example of plot and plot groups within each site; a,b,c etc. = individual plots within each site, a+a+a = plot group, etc.

Table 10.1.a The main type of growing condition and the number of plots and plot groups within each site.

Site No.	Type of grow. condition	Plot/	Plot gr.	Site No.	Type of grow. condition	Plot/	Plot gr.
1 -	cultivating systems	72	9	8 -	type and amount of fertilization	80	10
2 -	cultivating systems	48	9	9 -	amount of nitrogen	36	9
-	amount of nitrogene	48	9				
3-5-	sorts of sugar beets	36	9	10 -	amount of phosphorus and potassium	44	11
6 -	amount of insecticides	96	6	11 -	amount of nitrogen	48	12
7 -	sorts of sugar beets	66	11	12 -	cultivating systems	16	8
				13 -	harvest dates	36	6

All field work was performed by the Sugar Company.

The field was located in very flat terrain and the soil type was homogeneous within the field.

10.2.2 Aerial photography

A vertical aerial photograph (1:4000) of the sugar beet field was taken by using an aircraft and a camera with a 152.21mm lens (Zeiss LMK 265005A). The flight was carried out by The National Land Survey of Sweden at a height of 600m at 11.55 on the 30th of June 1986.

The camera held Kodak Aerochrome Infrared Film 2443 with a Kodak Wratten Filter No. 12. This film is sensitive to visible and near infrared radiation to approximately 900nm.

The sky was free from clouds during the flight.

Complete cover had not yet occurred when the photograph was taken.

10.2.3 Image digitization

An Optronics C-4500 multicolour scanner/plotter at The National Defence Research Institute in Linköping was used and the original positive transparancy (1:4000) was scanned sequentially through various filters so that green (G), red (R), and near infrared (IR) data were separated (note: there is always an overlap in film layer sensitivity). The spot size used for digitization was 25μm which made a pixel size of approximately 0.1m. The Optronic system recorded film densities as one of 256 levels for each pixel and stored the digitized data on magnetic tapes (1600bpi). A computer printout from these magnetic tapes thus showed "scene brightness values" in a range from 0 to 255 with high digital numbers (DN) indicating high reflectance and low values indicating low reflectance values.

10.2.4 Image correction

The incoming solar energy is composed of sunlight and skylight. The intensity of these and the relative relationship between these two are dependent on weather condition, wavelength, latitude, season of year, time of day and local topography. The airlight contains no information about the ground properties and this light is visually manifested as a general increase in scene brightness and a decrease in image contrast.

When comparing images taken by different equipment or on different occasions, locations etc. or when comparing scene points within one image with varying illumination conditions (due to for instance the local topography) this must be accounted for. One way to consider different illumination or atmospheric conditions is to relate the information in the photograph to the absolute reflectance of the ground objects. Methods to measure reflectance photographically are discussed by Lillesand and Kiefer (1979), Curran (1980) and Curran et al. (1981).

The present study is monotemporal and no comparision of images taken under different illumination and atmospheric conditions was performed. The investigation was accomplished in a rather small (150x500m), limited, absolutely flat field which leads to homogeneous illumination conditions within the field. When studying integrated spectral data (i.e. IR/R or normalized difference, ND) the most appropriate approach is to compensate for changes in the spectral quality of the incident light by calculating the spectral reflectance. In the present work preparations for relating photographic data to reflectance were not carried out and accordingly reflectance was not calculated.

To control the geometric accuracy of the image and correct possible errors a grid net of control points was placed in the field prior to the flight. Geometric disturbances were not found, however, and no geometric correction was performed. To avoid edge effects the field was located to the middle of the image and covered in the x-direction approximately 50 per cent of the image and in the y-direction about 20 per cent.

10.2.5 Measurements of digital numbers and statistical analysis

At the Remote Sensing Laboratory, University of Lund, the mean DN of each plot (Fig. 10.1.c) was measured. An EBBA image processing system was used.

The relationship between DN and the sugar beet yield and plant number was studied by Pearson's product-moment correlation (r) analysis (t-test was performed). These studies were carried out according to Tab. 10.2 and included simple linear as well as non-linear relationships. Tab. 10.3 presents the relationships examined in this work.

Analyses of variance (the F-test) were carried out regarding the relationship between spectral data of various:

- sugar beet varieties (18 different varieties)

- groups with different amount of fertilization;
  a. high and low amount
  b. high, mean and low amount
  c. 0,20,40,60,80,100,120,130,140,160 kg of fertilization/ha

Table 10.2 Studies of the relationship between sugar beet yield and aerial photography were performed among:

1 - all plots within each site

2 - the plot groups within each site

3 - all plots within each cultivation or treatment group

4 - the plot groups within each cultivation or treatment group

5 - all plots in all sites

6 - between the plot groups in all sites
   - for single bands (G, R and IR)
   - for the IR/R ratio and ND (IR-R/IR+R)

Table 10.3 The relationships examined in this work.

	relation-ship	band	point in Tab. 10.2		relation-ship	band	point in Tab. 10.2
1	linear	G,R,IR	1	19	linear	IR/R,ND	1
2	linear	G,R,IR	2	20	linear	IR/R,ND	2
3	linear	G,R,IR	3	21	linear	IR/R,ND	3
4	linear	G,R,IR	4	22	linear	IR/R,ND	4
5	linear	G,R,IR	5	23	linear	IR/R,ND	5
6	linear	G,R,IR	6	24	linear	IR,R,ND	6
7	log band	G,R,IR	1	25	log band	IR/R,ND	1
8	log band	G,R,IR	2	26	log band	IR/R,ND	2
9	log band	G,R,IR	3	27	log band	IR/R,ND	3
10	log band	G,R,IR	4	28	log band	IR/R,ND	4
11	log band	G,R,IR	5	29	log band	IR/R,ND	5
12	log band	G,R,IR	6	30	log band	IR/R,ND	6
14	log	G,R,IR	2	31	log		
15	band	G,R,IR	3	32	band		
16	and	G,R,IR	4	33	and		
17	log	G,R,IR	5	34	log		
18	yield	G,R,IR	6	35	yield		

- groups with different amount of insecticides;
  a. high and low amount.

The fertilization and insecticide treatment were performed before the air photograph was taken.

Statistically significant differences and analyses refer to the 95% level. The software package MINITAB (Ryan et al. 1985) was used for the statistical analysis.

Table 10.4.a The relationship between yield and single bands for all plots within each site (example 1 of Tab. 10.3).

Site	Band	No. of pl.	r	Site	Band	No. of pl.	r
1	IR	72	+0.24*	8	IR	80	+0.18
	R		-0.27*		R		-0.16
	G		-0.33*		G		-0.25
2	IR	48	+0.11	9	IR	36	+0.24
	R		-0.12		R		-0.04
	G		-0.15		G		0.0
3	IR	60	+0.50	10	IR	44	+0.30
	R		-0.52*		R		-0.06
	G		-0.52*		G		-0.08
4	IR	36	+0.38*	11	IR	48	+0.56*
	R		-0.17		R		-0.36*
	G		-0.12		G		-0.35*
5	IR	36	+0.45*	12	IR	16	+0.79*
	R		-0.42*		R		-0.79*
	G		-0.35*		G		-0.76*
6	IR	96	+0.43*	13	IR	36	+0.26
	R		-0.31*		R		-0.41*
	G		-0.36*		G		-0.43*
7	IR	66	+0.11				
	R		-0.06				
	G		-0.16				

No. of pl.= number of plots
r= the correlation coefficient
*= statistically significant at the 95% level

10.3 RESULTS

10.3.1 The relationship between sugar beet yield and digital numbers

Studies of the relationship between sugar beet yield and digitized aerial photography were performed on a single band basis as well as on a ratio (IR/R) and normalized difference basis.

Results are presented in Tabs. 10.4.a-e, Tabs. 10.5.a-e and Tab. 10.6. These tables give the correlation coefficient (r) and the statistical significance of the correlation analyses for examples 1-6 and 19-24 of Tab. 10.3. Scatter diagrams of the relationship between DN and sugar beet yield are presented by Hall-Könyves and Jiaju (1987).

Table 10.4.b The relationship between yield and single bands for the mean value of each plot group (Fig. 10.1.c) within each site (Fig. 10.1.b), (example 2 of Tab. 10.3).

Site	Band	No. of pl. gr.	r	Site	Band	No. of pl. gr.	r
1	IR	9	+0.41	8	IR	10	+0.28
	R		-0.59		R		-0.54
	G		-0.62		G		-0.40
2	IR	9	+0.16	9	IR	9	+0.32
	R		-0.91*		R		-0.07
	G		-0.79*		G		0.09
3	IR	15	+0.25	10	IR	11	+0.56
	R		-0.63*		R		-0.44
	G		-0.61*		G		-0.35
4	IR	9	+0.42	11	IR	12	+0.82*
	R		-0.34		R		-0.87*
	G		-0.35		G		-0.85*
5	IR	9	+0.43	12	IR	16	+0.69
	R		0.0		R		-0.78*
	G		0.0		G		-0.83*
6	IR	6	+0.74*	13	IR	6	+0.51
	R		-0.70		R		-0.08
	G		-0.82*		G		-0.69
7	IR	11	+0.19				
	R		0.0				
	G		-0.10				

No. of pl. gr. = number of plot groups
r= the correlation coefficient
*= statistically significant at the 95% level

Table 10.4.c The relationship between yield and single bands for all plots within each cultivation group (example 3 of Tabl. 10.3).

Cu. gr.	Band	No. of pl.	r	Cu. gr.	Band	No. of pl.	r
1	IR	96	+0.20	4	IR	272	+0.03
	R		0.0		R		-0.22
	G		0.0		G		-0.21
2	IR	120	0.0	5	IR	9	+0.28
	R		0.03		R		-0.05
	G		0.0		G		-0.03
3	IR	198	+0.20*	6	IR	11	+0.69*
	R		-0.44*		R		-0.77*
	G		-0.45*		G		-0.79*

No. of pl. = number of plots
Cu.gr.= cultivation group
r= the correlation coefficient
*= statistically significant at the 95% level

Table 10.4.d The relationship between yield and single bands for the mean value of each plot group within each cultivation group (example 4 of Tab. 10.3).

Cu. gr.	Band	No. of pl. gr.	r	Cu. gr.	Band	No. of pl. gr.	r
1	IR	21	+0.17	4	IR	47	+0.12
	R		0.0		R		-0.36
	G		0.0		G		-0.34
2	IR	18	0.19	5	IR	16	+0.45
	R		-0.21		R		-0.27
	G		-0.23		G		-0.29
3	IR	44	+0.18	6	IR	11	+0.93*
	R		-0.57*		R		-0.98*
	G		-0.56*		G		-0.98*

No. of pl. gr.= number of plot groups
Cu.gr.= cultivation group
r= the correlation coefficient
*= statistically significant at the 95% level

Table 10.4.e The relationship between yield and single bands for (i) all plots in all sites (example 5 Tab. 10.3) and (ii) the mean value of each plot group in all sites (example 6 Tab. 10.3).

Band	No. of pl.	r	Band	No. of pl. gr.	r
IR	674	+0.05	IR	124	+0.11
R		-0.13	R		-0.13
G		-0.11	G		-0.11

No. of pl. gr.= number of plot groups
No. of pl.= number of plots
r= the correlation coefficient
*= statistically significant at the 95% level

Table 10.5.a The relationship between yield and combined bands for all plots within each site (example 19 Tab. 10.3).

Site	Band	No. of pl.	r	Site	Band	No. of pl.	r
1	IR/R	72	+0.28*	8	IR/R	80	+0.04
	ND		+0.27*		ND		+0.04
2	IR/R	48	+0.17	9	IR/R	36	+0.17
	ND		+0.18		ND		+0.17
3	IR/R	60	+0.54*	10	IR/R	44	+0.26
	ND		+0.54*		ND		+0.26
4	IR/R	36	+0.44*	11	IR/R	48	+0.49*
	ND		+0.44*		ND		+0.52*
5	IR/R	36	+0.54*	12	IR/R	16	+0.82*
	ND		+0.54*		ND		+0.83*
6	IR/R	96	+0.50*	13	IR/R	36	+0.39*
	ND		+0.51*		ND		+0.40*
7	IR/R	66	+0.11				
	ND		+0.11				

No. of pl.= number of plots
r= the correlation coefficient
*= statistically significant at the 95% level

- Single bands.

The single band analysis of all plots and plot groups within each site (Tabs. 10.4.a-b) resulted in high correlations for site 3 (varieties of sugar beets), 11 (amount of nitrogen) and 12 (cultivating systems).

Division of sites into main cultivating groups (Tabs. 10.4.c-d) did not lead to increased correlation coefficients compared with the "within each site" approach (Tabs. 10.4.a-b). The cultivating groups 3 (variety of sugar beets) and 6 (cultivating systems) yielded however very strong correlations.

As can be seen from Tab. 10.4.e analyses of all plots together in all the 13 sites gave very low correlations.

- IR/R and ND.

When applying band combinations increased correlations (compared with the single band analyses) were generally obtained.

Tabs. 10.5.a-b show several sites with high correlations (site 3,5 - variety of sugar beets, 6 - amount of insecticides, 11 - amount of nitrogen and 12 - cultivating systems). The correlation coefficients of Tabs. 10.5.c-d (for plot and plot groups within each cultivation group) gave correlation coefficients approximately around 0.5.

Tab. 10.5.e presents the correlations of plots and plot groups in all sites together.

The number of cases with $/r/ > 0.5$ and statistically significant r (the 95% level) are summarized in Tab. 10.6. Application of logaritmic equations did not increase the relationships.

Thus, for some sites a close connection between sugar beet yield and air photography data was found. As can be seen from Tab. 10.1.b the distribution of yield within each site was rather similar and therefore cannot explain the variation of correlation coefficients that was obtained between sites.

10.3.2 The relationship between number of plants and digital numbers

The plant number of sites 4,5,7 and 12 were calculated on the same day as the photograph was taken. The correlation coefficient between the plant number and DN for each site was generally high and single band studies gave correlations for IR around approximately (+0.4)-(+0.7) and for R and G around (-0.5).

Table 10.5.b The relationship between yield and combined bands for the mean value of each plot group within each site (example 20 Tab. 10.3).

Site	Band	No. of pl. gr.	r	Site	Band	No. of pl. gr.	r
1	IR/R	9	+0.66*	8	IR/R	10	+0.46
	ND		+0.63*		ND		+0.47
2	IR/R	9	+0.90*	9	IR/R	9	+0.27
	ND		+0.90*		ND		+0.28
3	IR/R	15	+0.60*	10	IR/R	11	+0.03
	ND		+0.62*		ND		+0.03
4	IR/R	9	+0.07	11	IR/R	12	+0.89*
	ND		+0.06		ND		+0.90*
5	IR/R	9	+0.43	12	IR/R	8	+0.75*
	ND		+0.40		ND		+0.76*
6	IR/R	6	+0.96*	13	IR/R	6	+0.46
	ND		+0.96*		ND		+0.46
7	IR/R	11	+0.13				
	ND		+0.13				

No. of pl. gr.= number of plot groups
r= the correlation coefficient
*= statistically significant at the 95% level

Table 10.5.c The relationship between yield and combined bands for all test plots within each cultivation group (example 21 Tab. 10.3).

Cu. gr.	Band	No. of pl.	r	Cu. gr.	Band	No. of pl.	r
1	IR/R	96	+0.41*	4	IR/R	272	+0.27
	ND		+0.41*		ND		-0.29
2	IR/R	120	0.0	5	IR/R	176	+0.13
	ND		0.0		ND		+0.15
3	IR/R	198	+0.53*	6	IR/R	64	+0.46*
	ND		+0.52*		ND		+0.45*

No. of pl. = number of plots
Cu.gr.= cultivation group
r= the correlation coefficient
*= statistically significant at the 95% level

Table 10.5.d The relationship between yield and combined bands for the mean value of each plot group within each cultivation group (example 22 Tab. 10.3).

Cu. gr.	Band	No. of pl. gr.	r	Cu. gr.	Band	No. of pl. gr.	r
1	IR/R	18	+0.44*	4	IR/R	47	+0.36
	ND		+0.45*		ND		+0.39
2	IR/R	44	+0.17	5	IR/R	16	+0.33
	ND		+0.18		ND		+0.34
3	IR/R	47	+0.63*	6	IR/R	17	+0.63*
	ND		+0.62*		ND		+0.60*

No. of pl. gr. = number of plot groups
Cu.gr.= cultivation group
r= the correlation coefficient
*= statistically significant at the 95% level

Table 10.5.e The relationship between yield and combined bands for (i) all plots within all sites (example 23 Tab. 10.3) and (ii) the mean of each plot group within all sites (example 24 Tab. 10.3).

Band comb.	No. of pl.	r	Band comb.	No. of pl. gr.	r
IR/R	674	+0.12	IR/R	124	+0.07
R		+0.13	R		+0.08

No. of pl.= number of plots
No. of pl. gr.= number of plot groups
r= the correlation coefficient

Table 10.6 The number of situations with /r/ > 0.5 and statistically significant (the 95% level).

Point in Tab. 10.2	Total number of sites or groups	G	R	IR	IR /R	ND
1	13	2	2	2	4	5
2	13	5	4	2	6	6
3	6	1	1	1	1	1
4	6	2	2	1	2	2
5	1	0	0	0	0	0
6	1	0	0	0	0	0

```
ANALYSIS OF VARIANCE
SOURCE      DF        SS         MS          F
FACTOR      17  0.019963   0.001174       4.94
ERROR      180  0.042774   0.000238
TOTAL      197  0.062737
                                          INDIVIDUAL 95 PCT CI'S FOR MEAN
                                          BASED ON POOLED STDEV
LEVEL        N      MEAN      STDEV       ---------+---------+---------+-------
S-1-ND      26   0.71658    0.01668                        (---*---)
S-2-ND      26   0.72670    0.01718                               (---*---)
S-3-ND      18   0.73279    0.01842                                  (----*-----)
S-4-ND      18   0.72689    0.01737                              (----*----)
S-5-ND      18   0.70875    0.01309                 (----*----)
S-6-ND       8   0.70613    0.01520             (------*-------)
S-7-ND       8   0.70396    0.01887           (-------*-------)
S-8-ND       8   0.70940    0.01170               (-------*------)
S-9-ND       8   0.70111    0.01197         (-------*------)
S10-ND       8   0.70769    0.01288              (------*-------)
S11-ND       8   0.69939    0.01450        (-------*------)
S12-ND       8   0.70296    0.01602          (-------*-------)
S13-ND       6   0.71506    0.00662                 (---------*--------)
S14-ND       6   0.71347    0.01504                (--------*--------)
S15-ND       6   0.71124    0.00999              (--------*---------)
S16-ND       6   0.71916    0.01182                    (--------*---------)
S17-ND       6   0.72422    0.01386                       (---------*---------)
S18-ND       6   0.71717    0.01115                  (---------*--------)
                                          ---------+---------+---------+-------
POOLED STDEV =  0.01542                          0.700     0.714     0.728


S-1 - S-18= various sorts of sugar beets
ND= normalized difference
```

Figure 10.2 The mean ND of various sugar beet sorts.

10.3.3 The relationship between various sugar beet varieties, various amounts of fertilization and insecticides and digital numbers

Analyses of variance between DN (G, R, IR, IR/R and ND) of various sugar beet varieties, fertilization and insecticide amounts produced in some cases statistically significant results.

```
ANALYSIS OF VARIANCE
SOURCE     DF        SS          MS           F
FACTOR      2   0.004911    0.002455      22.01
ERROR      93   0.010377    0.000112
TOTAL      95   0.015288
                                      INDIVIDUAL 95 PCT CI'S FOR MEAN
                                      BASED ON POOLED STDEV
LEVEL       N      MEAN      STDEV   --+---------+---------+---------+----
N-IN-ND    32   0.74420    0.01169                       (----*----)
H-IN-ND    32   0.74551    0.01143                         (----*----)
L-IN-ND    32   0.72972    0.00821   (----*----)
                                     --+---------+---------+---------+----
POOLED STDEV =  0.01056              0.7275    0.7350    0.7425    0.7500

N-IN= normal amount of insecticide
H-IN= high amount of insecticide
L-IN= low amount of insecticide
ND= normalized difference
```

```
ANALYSIS OF VARIANCE
SOURCE     DF        SS          MS           F
FACTOR     10   0.023363    0.002336       6.03
ERROR     165   0.063877    0.000387
TOTAL     175   0.087240
                                      INDIVIDUAL 95 PCT CI'S FOR MEAN
                                      BASED ON POOLED STDEV
LEVEL       N      MEAN      STDEV   ----------+---------+---------+------
F0-ND      12   0.71129    0.02770   (-----*----)
F20-ND      8   0.74433    0.01059                  (------*------)
F40-ND     16   0.74235    0.01332                   (----*----)
F60-ND     12   0.74407    0.01144                   (-----*-----)
F80-ND     16   0.74974    0.01557                       (----*----)
F100-ND    20   0.72703    0.01625            (----*---)
F120-ND    44   0.75285    0.02442                           (--*--)
F130-ND    16   0.73276    0.01616               (---*----)
F140-ND     4   0.74338    0.01443               (---------*--------)
F160-ND    24   0.74330    0.02239                     (---*---)
F180-ND     4   0.73977    0.00790             (---------*---------)
                                     ----------+---------+---------+------
POOLED STDEV =  0.01968                      0.720     0.740     0.760

F0= no fertilization (nitrogene)
F20 - F180= fertilization (nitrogene) 20 - 180 kg/ha
ND= normalized difference
```

Figure 10.3 The mean ND of plots with low, normal and high amount of insecticides and the mean ND of plots with various amount of fertilization.

The mean ND of various varieties is presented in Fig. 10.2 and the ND of plots with various amount of fertilization in

10.3. The plots with no fertilization were found to have the lowest ND. Fig. 10.3 also shows the mean ND of plots with low, normal and high amounts of insecticides. It can be seen that the lowest amount of insecticide is related to the lowest ND but that the ND of normal and high amounts are rather similar. These results thus indicate that spectral data may be sensitive to various varieties of sugar beets, fertilization amounts and insecticide amounts. However, statistical significance could not be obtained in this study.

10.4 STUDY REMARKS

* Spectral reflectance was not calculated.

* Yield data and plant number are valid for two rows within each plot while digital numbers are valid for the whole plot.

* GLAI was not estimated.

* Remote sensing of sugar beet yield is an indirect estimation of subterranean plant parts from above ground plant mass.

* The relationship between early season GLAI and at harvest root sugar mass is not straightforward (these complex relationships must be very well understood).

10.5 SUMMARY

The aim of the work presented in this section was to examine the relationship between digitized colour infrared aerial photography and various sugar beet variables like yield, plant number, variety and amount of fertilization and insecticides.

The sugar beet variables plant density and variety, as well as the amount of fertilization and amount of insecticides affected the DN of the aerial photograph while the relationship between DN and sugar beet yield was ambiguous.

11

GENERAL SUMMARY

The aim of the study was to investigate the potential for remote sensing application in Swedish agriculture with special reference to satellite-based monitoring.

The first part of the work described preprocessing approaches including relative calibration, examination of various sun angle and atmospheric correction methods, and a study of the topographic effect. The second part of the thesis concentrated mainly on crop classification and crop area estimation. In addition, the relationships between remotely-sensed data and crop types, crop varieties, crop yield and various amounts of insecticides and fertilization were studied.

Relative calibration was performed by calculating in-band radiance (mW/(cm² sr)) and spectral radiance (mW/(cm² sr μm)).

In order to determine the most efficient approach to reduce between-scene variability, the coefficient of variation was calculated for various transformation/band/"unchanging" surface combinations. Two approaches were found to be suitable, namely the effective at-satellite reflectance and the method that considers the optical air mass and the atmospheric transmission constant.

The between-scene variability caused by atmospheric influence was reduced through (i) the traditional haze removal and (ii) a standardization approach.

Digital elevation data were merged with digital satellite data in order to evaluate the topographic effect of a gently undulating terrain on LANDSAT data. It was concluded that the topographic conditions have limited importance for crop classification in the south of Sweden.

Monotemporal LANDSAT data of crop type differences produced statistically significant differences between the winter crop group and the summer crop group at all studied acquisition dates. This also applied to rape seed, potatoes and sugar beet, while individual cereals seldom produced any differences. In July, however, rye differed from summer wheat, barley and oats. The classification accuracy of this crop was rather high as well. The appearances of various multitemporal crop profiles were strongly influenced by the application of the solar angle correction methods. Various solar angle corrected temporal-spectral profiles were generally associated with an "overcompensation" for low sun angles.

Low classification accuracies were generally noted. Rye and potatoes produced the highest values, obtaining Kappa coefficients around 70%. The classification results did not appear to have been influenced by soil type variations. The accuracy, however, was dependent on the sizes of the fields and the preprocessing approach used. It was not proved that multitemporal classification is superior to monotemporal classification.

Area estimations of winter wheat, rye, winter rape-seed, sugar beet, and barley were performed. The overestimation for rye and winter wheat was very large (over 100%) while the correspondence between actual and estimated area was relatively good for barley and very good for winter rape-seed and sugar beet. The results thus indicated a high potential for large area estimation for certain crops in the south of Sweden.

Aerial photography data produced statistically significant differences between various sugar beet varieties, indicating the possible influence of variety variation on large area sugar beet studies.

Fertilization and insecticides caused effects on sugar beet plots that affected the digital photography. Thus a connection between the amount of fertilization and insecticides and the digital numbers was revealed.

The applicability of LANDSAT MSS data for crop yield was low for the majority of crops grown in the south of Sweden. Nevertheless, statistically significant correlations were obtained for winter wheat, barley, and sugar beet. Analyses of the correlation between LANDSAT TM data and biological yield of various crops showed that winter wheat often produced the strongest relationships. Thus this crop seems to be promising for production assessment. The connection between late season (September) TM data and sugar beet yield was very low. This was also the case with the relation between early season (June) aerial photography data and sugar beet yield. However, the relationship between these data and plant density was rather strong.

Observations were made which clearly indicate that remote sensing methods can be of great use in some Swedish agricultural applications. The limitations of the applicability of the methods used in this work have been pointed out. Further research will show if the methods can be improved and the application of remote sensing data successfully extended in Swedish agriculture.

APPENDIX A

DIFFERENCES BETWEEN CROP TYPES

Y= Statistically significant differences between crop types (the 95% level)
N= Not statistically significant differences between crop types (the 95% level).
Area= Sub-areas 1-6 (Fig. 4.1).

AREA 1

	RYE
TM:	345744547 37753
W.WHEAT	
JUNE 07	NYYYYYYYN
JULY 09	YYYYYYYYY

AREA 2

	S.WHEAT	RYE	BARLEY	OATS	POT.	GR.CR.	RAPE S.
TM:	345744547 37753	345744547 37753	345744547 37753	345744547 37753	345744547 37753	345744547 37753	345744547 37753
W.WHEAT							
JUNE	NNYNNNNN	NYYYYYYYN	YYYNYYYYN	NNNNNNNNN	YYYYYYYYY	YYNYYYNYN	YYYYYYYYY
JULY	NYYYYYYN	YYNYYYYYN	NYYNYYYYN	NYNYNNNYN	YYYYYYYYY	YNYYYYNYN	YYYYYYYYY
S.WHEAT							
JUNE		NYYYYYYYN	NYNNYYNYN	NNYNNNNN	YYYYYYYYY	YYNYYYNYN	YYYYYYYYY
JULY		YYYYYYYYY	NNNNNNNNN	YYYYYYYYY	YYYYYYYYY	YNYYYYYYN	YYYYYYYYY
RYE							
JUNE			YYYYYYYYY	NYYYYYNYN	YYYYYYYYN	NYYNNNNYN	YYYYYYYYY
JULY			YYYYYYYYY	YYNYYYNYN	YYYYYYYYY	NYYYYYYYY	YYYYYYYYY
BARLEY							
JUNE				NYYNYYNYN	YYYYYYYYY	YYNNYYNYN	YYYYYYYYY
JULY				YNYYYYYYN	YYYYYYYYY	YYYYYYYYY	YNYYYYYYY
OATS							
JUNE					YYYYYYYY	YYNNYNYYN	YYYYYYYYY
JULY					YYYYYYYY	YYYYYYYYY	YNYYNNYNN
POT.							
JUNE						YYYYYYYYY	NNNNNNNNN
JULY						YYYYYYYYY	YNNYNNYNN
GR.CR.							
JUNE							YYYYYYYYYY
JULY							YYYYYYYYYY

AREA 3

	S.WHEAT	RYE	BARLEY	OATS	POT.	GR.CR.	RAPE S.
TM:	345744547 / 37753	345744547 / 37753	345744547 / 37753	345744547 / 37753	345744547 / 37753	345744547 / 37753	345744547 / 37753
W.WHEAT							
JUNE	NYYYYYYYN	YYNYYYYYN	NYNNNYYNN	NYNNNYYYN	YYYYYYYYY	YNYYYYNYN	YYYYYYYYY
JULY	NYYNNYYYN	NNNNNNNNN	NNYYNNNNN	NNNNNNNNN	YYYYYYYYY	NYYNYYNYN	YYYYYYYYY
S.WHEAT							
JUNE		NYYNYYNYN	NYNNYYNYN	NYNNYYNNN	YYYYYYYYY	NNNNNNNNN	YYYYYYYYY
JULY		YYYYYYYYY	NNNNNNNNN	NNNYNNYNN	YNYYYYNNN	YYYYYYYYY	YYYYYYYYY
RYE							
JUNE			NYYNNYYNN	NYNNNNNNN	YYYYYYYYY	NYYNYYNNN	YYYYYYYYY
JULY			YYYYYYYYY	YYNYYYNNN	NYYYYYYYN	YNYYYYNNN	YYYYYYYYY
BARLEY							
JUNE				NNNNNNNNN	YYYYYYYYY	NNNNNNNNN	YYYYYYYYY
JULY				NYNNYYNNN	YNYYNNYYN	YYYYYYYYY	YYYYYYYYY
OATS							
JUNE					YYYYYYYYY	NNNNNNNNN	YYYYYYYYY
JULY					NNYYYYNNN	YYYYYYYYY	YYYYYYYYY
POT.							
JUNE						YYYYYYYYY	YYYYYYYYY
JULY						YYYNYYNNN	YYYYYYYYY
GR.CR.							
JUNE							YYYYYYYYY
JULY							YYYYYYYYY

AREA 4

	S.WHEAT	RYE	BARLEY	OATS	POT.	GR.CR.	RAPE S.
TM:	345744547 / 37753	345744547 / 37753	345744547 / 37753	345744547 / 37753	345744547 / 37753	345744547 / 37753	345744547 / 37753
W.WHEAT							
JUNE		YYYYYYYYY	NYNYYYNNN	NYNNYNNNN	X	NYNNNNNNN	X
JULY		YNNNYYNNN	NYNYYYNNN	YYNNYNNNN	X	YNYYYYNNN	X
S.WHEAT							
JUNE		X	X	X	X	X	X
JULY		X	X	X	X	X	X
RYE							
JUNE			NYYYYYYYN	NYYYYYYYY	X	NYYYYYYYYN	X
JULY			NYYYYYYYN	NYYYYYYYN	X	NNNNNNNNNN	X
BARLEY							
JUNE				NYNYYYNNN	X	NNNNNNNNNN	X
JULY				NNNNNNNNN	X	YNYYYYNNNN	X
OATS							
JUNE					X	NNNNNNNNNN	X
JULY					X	NNNNNNNNNN	X
POT.							
JUNE						X	X
JULY						X	X
GR.CR.							
JUNE							X
JULY							X

AREA 5

	S.WHEAT	RYE	BARLEY	OATS	POT.	GR.CR.	RAPE S.
TM:	345744547	345744547	345744547	345744547	345744547	345744547	345744547
	37753	37753	37753	37753	37753	37753	37753
W.WHEAT							
JUNE	NNNNNNNNN	NYYYYYNNN	YYYNYYNNN	YNNNYYNNN	YYYYYYYYY	NYNNYYNYN	YYYYYYYYY
JULY	NYYNYYNNN	YYNNYYNNN	NYYNYYNYN	NYYNYYNYN	YYYYYYYYY	NNNNNNNNN	YYYYYYYYY
S.WHEAT							
JUNE		NYNNYYNNN	NNNNNNNNN	NNNNNNNNN	YYYYYYYYY	NNNNNNNNN	YYYYYYYYY
JULY		NYNNYYNNN	NNNNNNNNN	NNNNNNNNN	YYYYYYYYY	NNNNNNNNN	YYYYYYYYY
RYE							
JUNE			NYYYYNNNN	YYYYYYYYY	YYYYYYYYY	NYYYYYNYN	YYYYYYYYY
JULY			YYNYYYNYN	YYYYYYYYY	YYYYYYYYY	YNYYYYNYN	YYYYYYYYY
BARLEY							
JUNE				YYNNYYNYN	YYYYYYYYY	NNNNNNNNN	YYYYYYYYY
JULY				NNNNNNNNN	YYYYYYYYY	YYYYYYYYY	NNYYYYNYN
OATS							
JUNE					YYYYYYYYY	YYNYYYNYN	YYYYYYYYY
JULY					NYYYYYNYN	YYYYYYYYY	YNYYYYYYY
POT.							
JUNE						YYYYYYYYY	NNNNNNNNN
JULY						YYYYYYYYY	NYNNNNNNN
GR.CR.							
JUNE							YYYYYYYYY
JULY							YYYYYYYYY

AREA 6

	S.WHEAT	RYE	BARLEY	OATS	POT.	GR.CR.	RAPE S.
TM:	345744547	345744547	345744547	345744547	345744547	345744547	345744547
	37753	37753	37753	37573	37753	37753	37753
W.WHEAT							
JUNE	NYNNYYNYN	NYYYYYNYN	YYNNYYNYN	YNNNYYNYN	X	YNYYYYNYN	X
JULY	NNNNNNNNN	YNNNNNNNN	NNNNNNNNN	NNNNNNNNN	X	YYYYYYYYY	X
S.WHEAT							
JUNE		NYYYYYNYN	YNNNNNNNN	NNNNNNNNN	X	YYYYYYYYY	X
JULY		YNNYYYNYN	NNNNNNNNN	YNNYYNNNN	X	YYYYYYYYY	X
RYE							
JUNE			YYYYYYYYY	NYYYYYNYN	X	YYYNYYNYN	X
JULY			YNNYYYNYN	YNNNYYNYN	X	YYYYYYYYY	X
BARLEY							
JUNE				NNNNNNNNN	X	YYYYYYYYY	X
JULY				NNNNNNNNN	X	YYYYYYYYY	X
OATS							
JUNE					X	YYYYYYYYY	X
JULY					X	YYYYYYYYY	X
POT.							
JUNE						X	X
JULY						X	X
GR.CR.							
JUNE							X
JULY							X

APPENDIX B

MSS DATA - AT-HARVEST YIELD

The Pearson product-moment correlation coefficient (r) between MSS data acquired at three dates and at-harvest grain yield (kg/ha).
1-30= reference to a scatter plot presented in this appendix.

		W.Wheat n=46	S.Wheat n=32	Rye n=33	Barley n=43	Oats n=35	Leg. n=31	Grazing crop, n=41	Rape seed, n=36	Sugar beet, n=20
MSS1:	May 19	-0.58	-0.28	0.45	0.56	-0.25	-0.45	-0.23	0.17	-0.01
	June 20	-0.26,1	-0.46	-0.53	0.40,11	-0.30	-0.42	-0.24	0.18	-0.08,21
	July 06	-0.36,2	-0.65	-0.36	0.55,12	-0.11	-0.21	0.13	0.08,22	
MSS2:	May 19	-0.57	-0.29	-0.12	-0.21	-0.23	-0.54	-0.31	0.05	0.16
	June 20	-0.32,3	-0.49	-0.34	-0.42,13	-0.28	-0.41	-0.22	0.08	0.16,23
	July 06	-0.57,4	-0.52	-0.13	-0.61,14	-0.31	-0.59	-0.29	0.19	-0.03,24
MSS3:	May 19	0.29	0.04	0.03	-0.49	-0.29	-0.27	-0.11	0.48	0.37
	June 20	0.32,5	0.16	0.49	0.30,15	0.25	0.37	0.24	0.22	0.47,25
	July 06	0.51,6	0.07	0.16	0.45,16	0.32	0.55	0.28	0.04	0.58,26
MSS4:	May 19	0.41	0.01	0.03	-0.51	-0.31	-0.25	-0.24	0.42	0.38
	June 20	0.51,7	0.17	0.33	0.26,17	0.39	0.38	0.15	0.17	0.54,27
	July 06	0.56,8,	0.16	0.12	0.57,18	0.36	0.59	0.23	-0.10	0.74,28
MSS4/2:	May 19	0.53	0.29	0.18	-0.23	-0.22	0.36	-0.19	0.09	0.23
	June 20	0.48,9	0.31	0.27	0.35,19	0.38	0.42	0.34	0.10	0.31,29
	July 06	0.68,10	0.40	0.07	0.67,20	0.35	0.59	0.31	-0.12	0.54,30
LogMSS1:	May 19	-0.57	-0.22	-0.46	-0.53	-0.27	-0.20	-0.14	0.17	-0.19
	June 20	-0.25	-0.42	-0.53	-0.35	-0.29	-0.29	-0.26	0.38	-0.12
	July 06	-0.35	-0.65	-0.36	-0.53	0.05	-0.31	-0.13	0.04	0.23
LogMSS2:	May 19	-0.57	-0.21	-0.19	-0.13	-0.30	-0.41	-0.11	0.09	0.16
	June 20	-0.32	-0.42	-0.35	-0.39	-0.37	-0.39	-0.09	0.11	0.09
	July 06	-0.57	-0.50	-0.12	-0.62	-0.36	-0.41	-0.12	0.05	-0.03
LogMSS3:	May 19	0.22	0.08	0.05	-0.43	-0.35	-0.21	-0.24	0.39	0.34
	June 20	0.32	0.15	0.48	0.41	0.37	0.35	0.11	0.18	0.45
	July 06	0.52	0.06	0.15	0.45	0.33	0.45	0.36	0.06	0.59
LogMSS4:	May 19	0.43	0.00	0.05	-0.51	-0.36	-0.15	-0.24	0.40	0.39
	June 20	0.51	0.27	0.32	0.33	0.31	0.29	0.21	0.19	0.53
	July 06	0.57	0.17	0.11	0.58	0.38	0.43	0.29	0.01	0.74

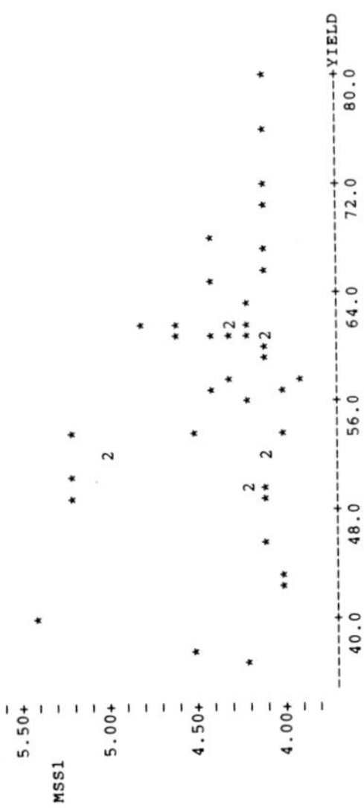
MSS1
5.50+
5.00+
4.50+
4.00+
40.0
48.0
56.0
64.0
72.0
80.0
YIELD
1

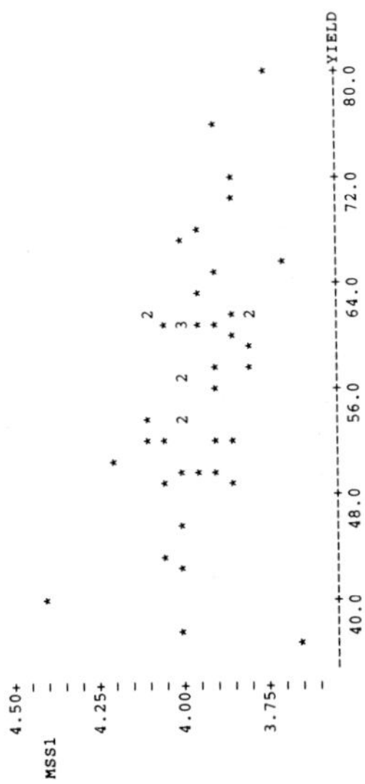
MSS1
4.50+
4.25+
4.00+
3.75+
40.0
48.0
56.0
64.0
72.0
80.0
YIELD
2

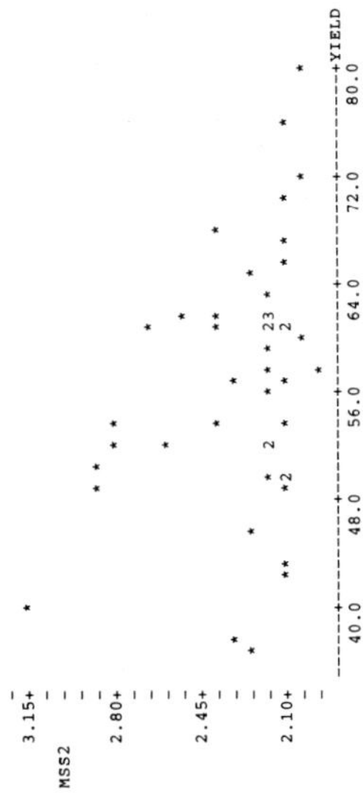
MSS2
3.15+
2.80+
2.45+
2.10+
40.0
48.0
56.0
64.0
72.0
80.0
YIELD
3

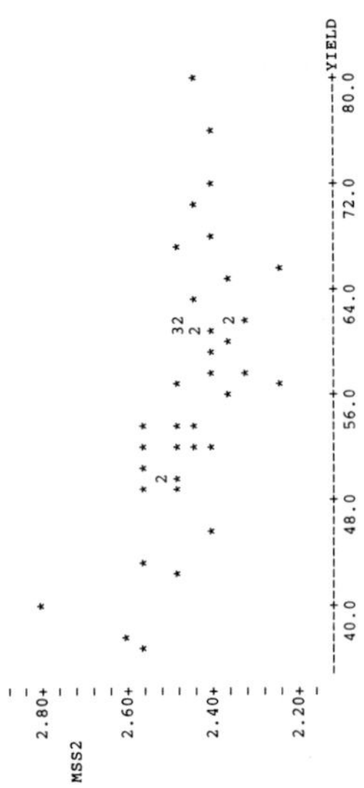
MSS2
2.80+
2.60+
2.40+
2.20+
40.0
48.0
56.0
64.0
72.0
80.0
YIELD
4

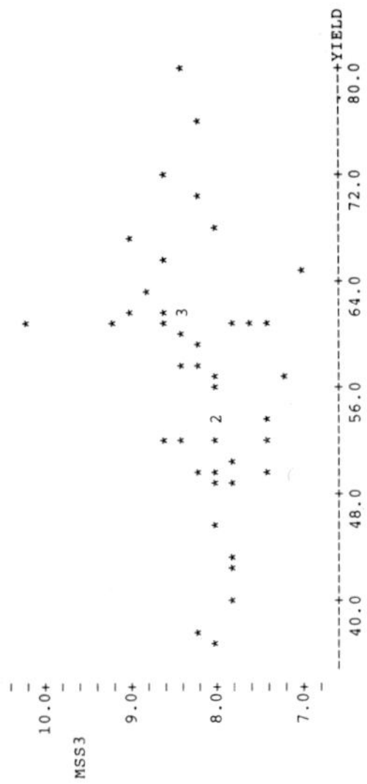
MSS3
10.0+
9.0+
8.0+
7.0+
40.0
48.0
56.0
64.0
72.0
80.0
YIELD
5

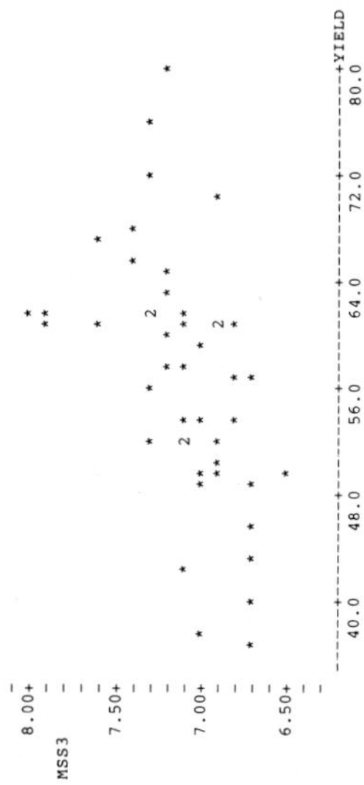
MSS3
8.00+
7.50+
7.00+
6.50+
40.0
48.0
56.0
64.0
72.0
80.0
YIELD
6

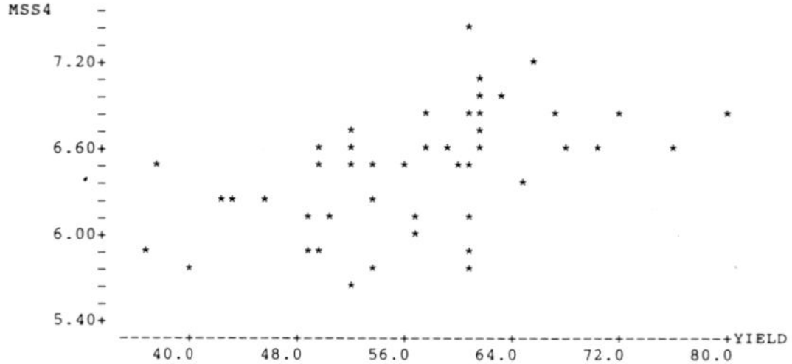
MSS4
7.20+
6.60+
6.00+
5.40+
40.0
48.0
56.0
64.0
72.0
80.0
YIELD
7

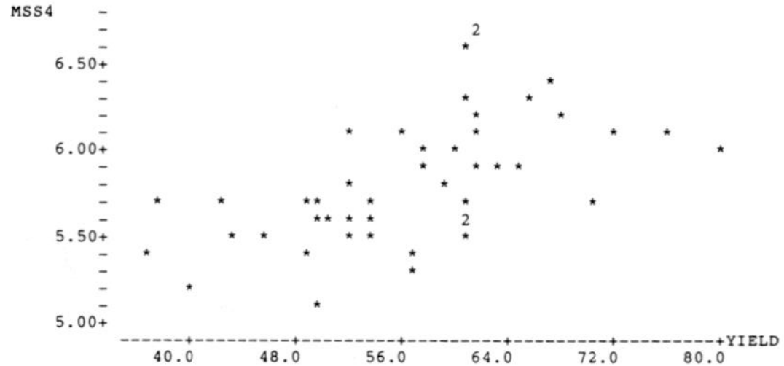
MSS4
6.50+
6.00+
5.50+
5.00+
40.0
48.0
56.0
64.0
72.0
80.0
YIELD
8

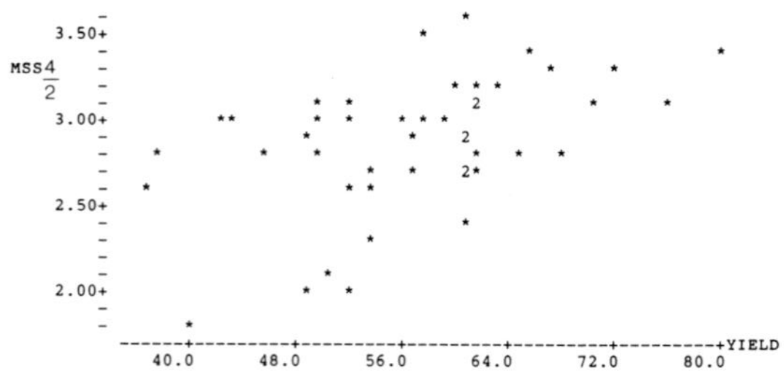
MSS4/2
3.50+
3.00+
2.50+
2.00+
40.0
48.0
56.0
64.0
72.0
80.0
YIELD
9

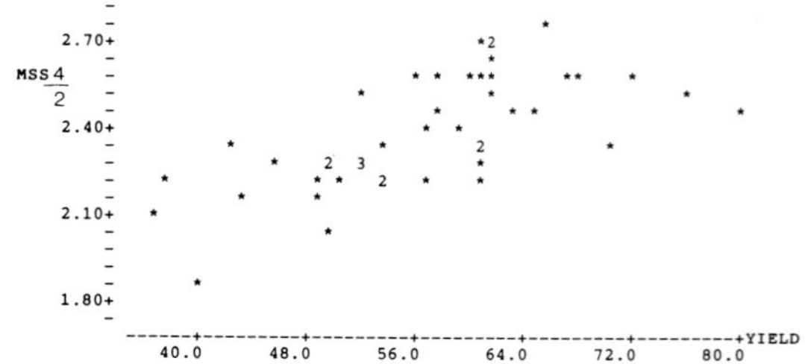
MSS4/2
2.70+
2.40+
2.10+
1.80+
40.0
48.0
56.0
64.0
72.0
80.0
YIELD
10

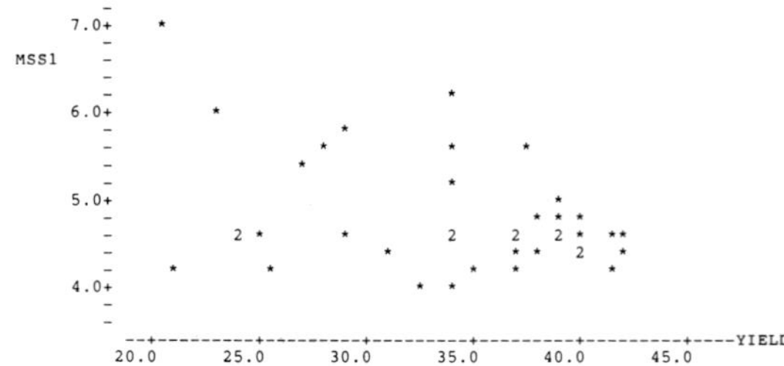
MSS1
7.0+
6.0+
5.0+
4.0+
20.0
25.0
30.0
35.0
40.0
45.0
YIELD
11

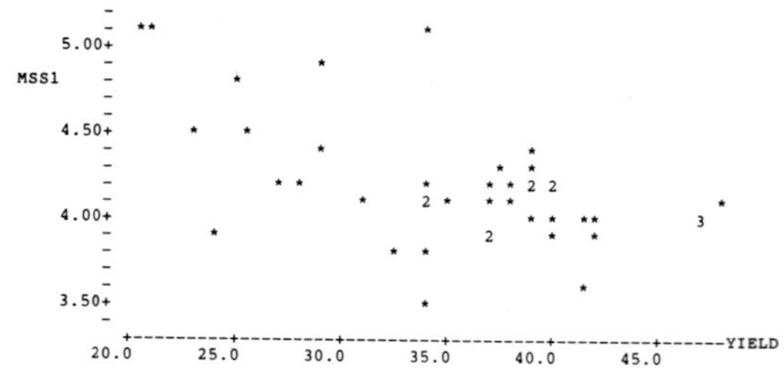
MSS1
5.00+
4.50+
4.00+
3.50+
20.0
25.0
30.0
35.0
40.0
45.0
YIELD
12

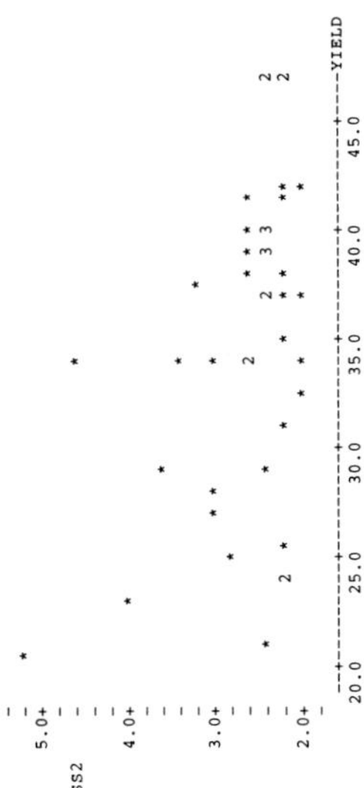

13

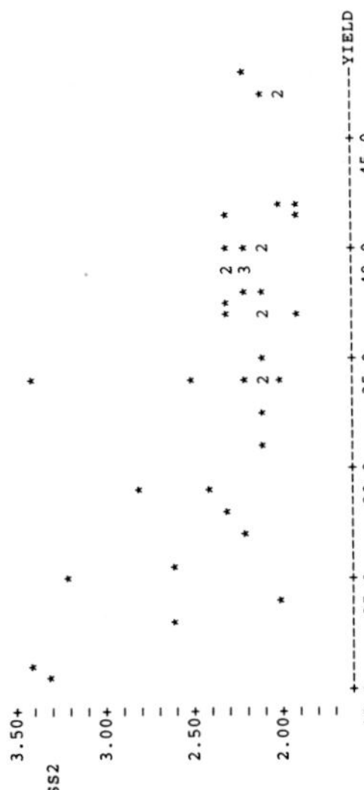

14

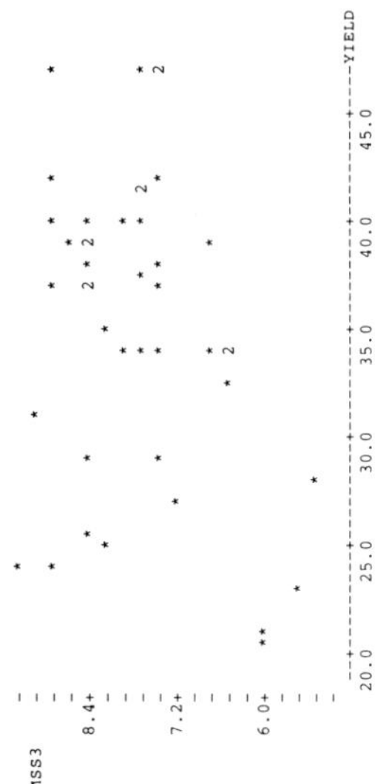

15

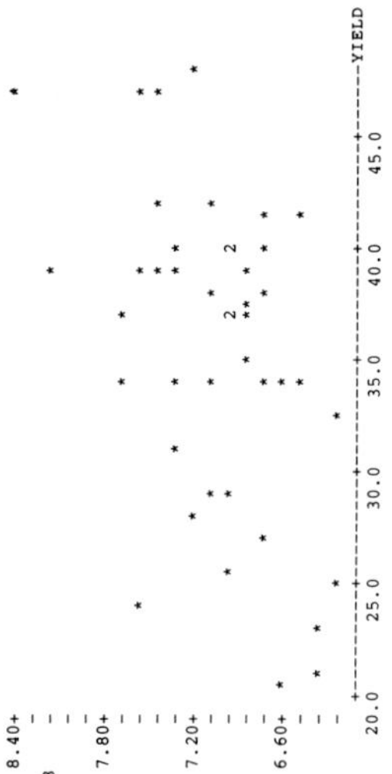

16

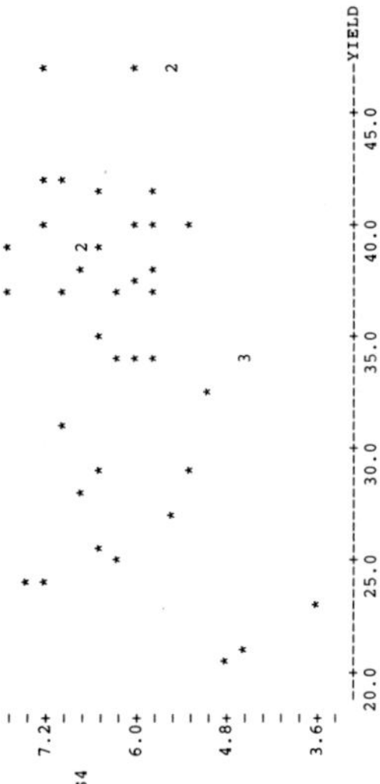

17

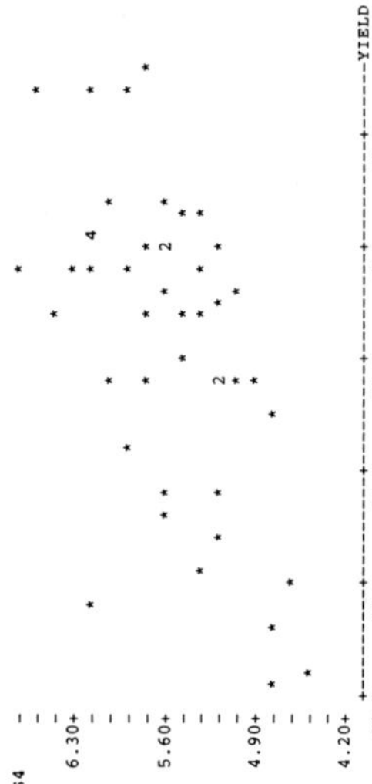

18

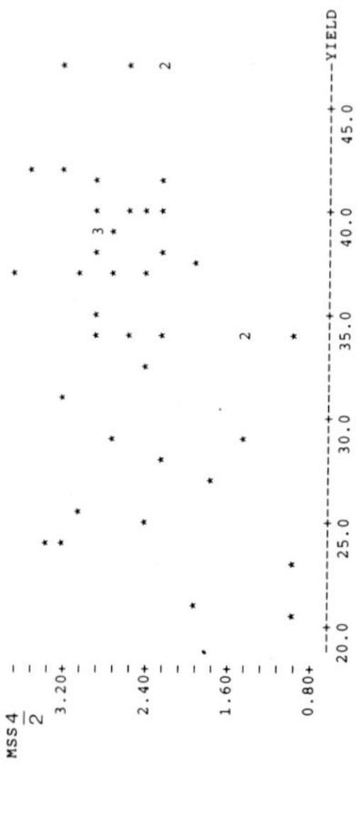

19

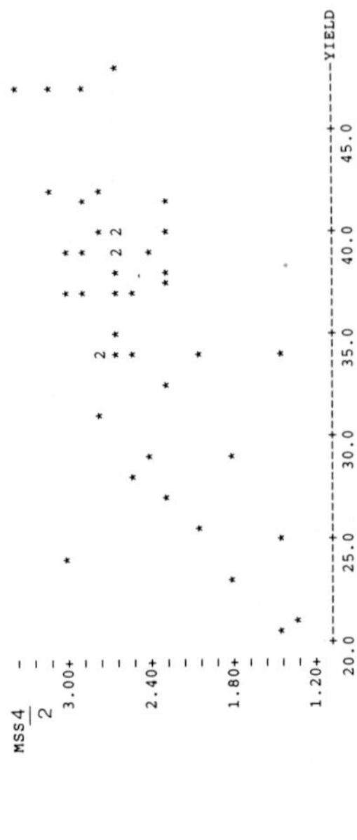

20

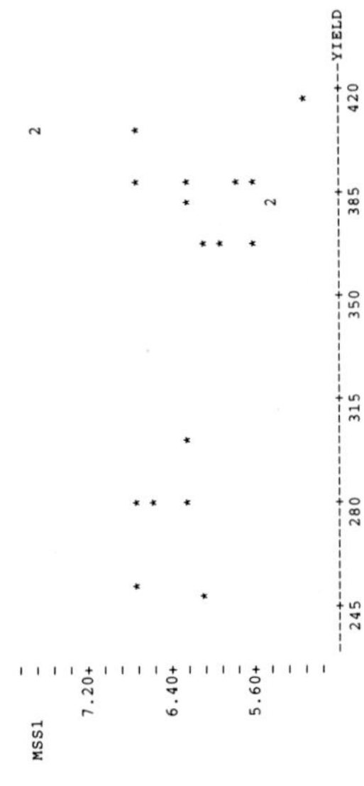

21

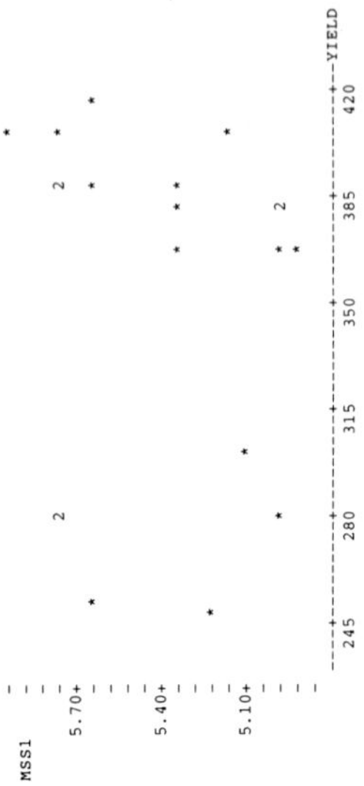

22

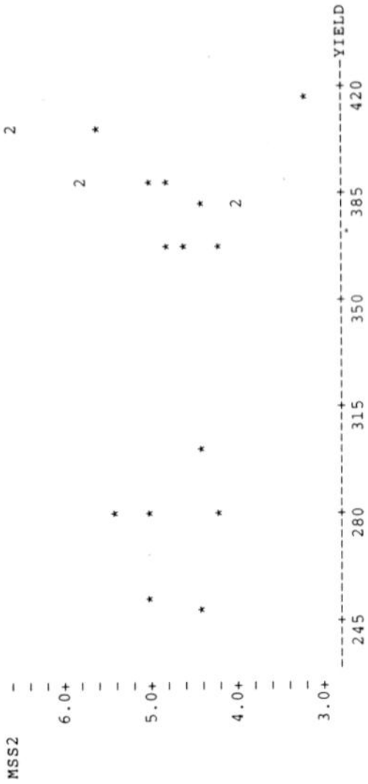

23

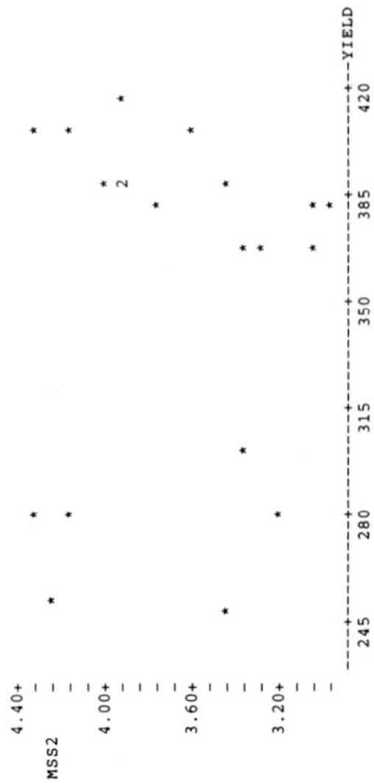

24

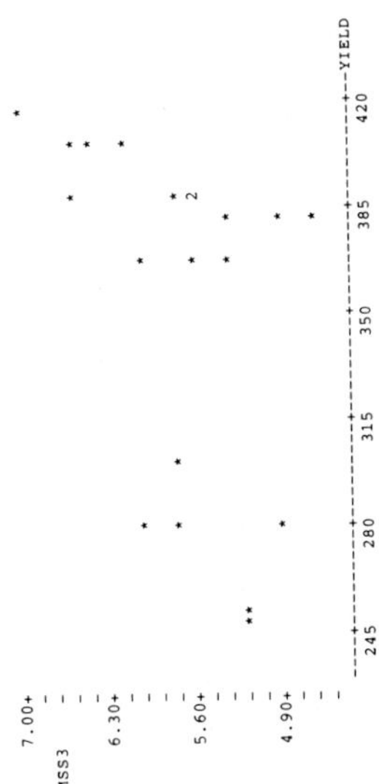

25

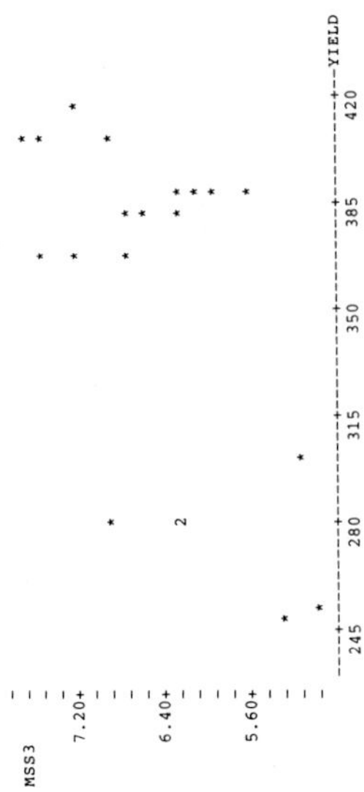

26

27

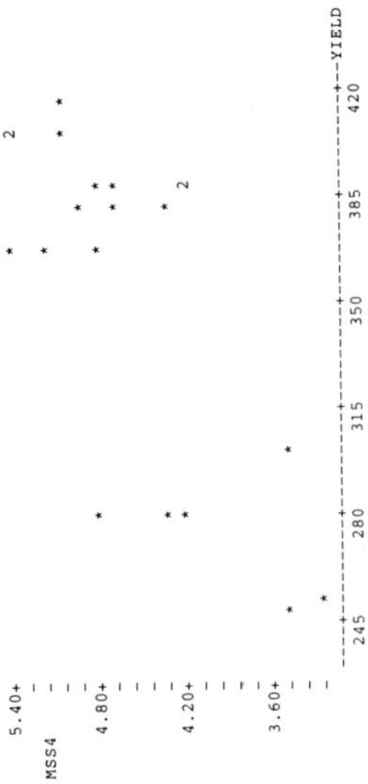

28

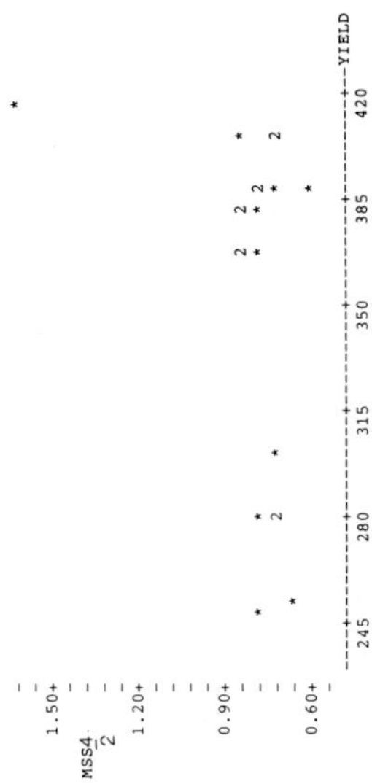

29

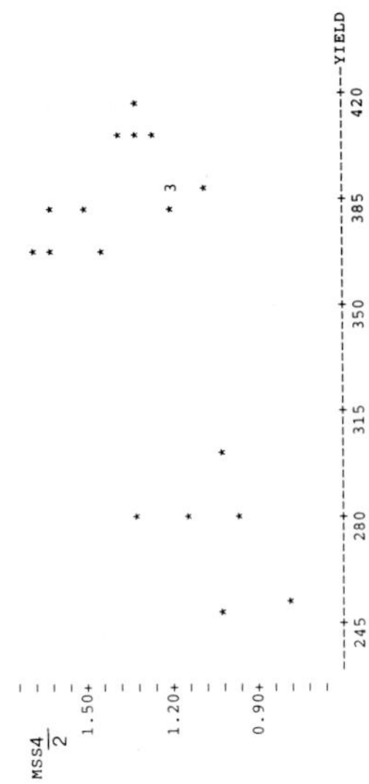

30

APPENDIX C

TM DATA - BIOLOGICAL YIELD

The Pearson product-moment correlation coefficient (r) between TM data and biological yield.

1-46= reference to a scatter plot presented in this appendix.
SSA1-SSA6= sub-areas 1-6 (Fig. 4.1).

	W.Wheat n=29	Rye n=23	W.Wheat n=29	Rye n=23
SSA1, JUNE			SSA1, JULY	
TM3	-0.26	-0.06	-0.06	-0.10
TM4	0.24,1	0.23,3	0.41	0.11
TM5	-0.54,2	-0.08	0.42	-0.13
TM7	-0.65	-0.04	-0.43,4	-0.19
TM4/3	0.25	0.21	0.26	0.05
TM7/3	-0.65	-0.09	-0.41,5	-0.06
TM4/7	0.55	0.22	0.42	0.09
TM5/3	-0.37	-0.66	-0.05	-0.05
ND	0.31	0.24	0.38	0.13
LOGTM4	-0.25	-0.05	-0.07	-0.11
LOGTM5	0.27	0.25	0.35	0.13
LOGTM5	-0.51	-0.06	-0.41	-0.09
LOGTM7	-0.62	-0.03	-0.39	-0.10

	W.Wheat n=110	S.Wheat n=26	Rye n=61	Barley n=108	Oats n=70	Potatoes n=46	Grazing crop, n=48
SSA2, JUNE							
TM3	-0.29	-0.15	-0.28	-0.05	-0.04	-0.17	-0.41
TM4	0.32,6	0.05	0.38 10	0.01,12	0.39,13	0.06	0.45
TM5	-0,46,	-0.17	-0.01	-0.16	-0.16,	-0.16	-0.34
TM7	-0.48,7	-0.13	-0.13	-0.13	-0.04	-0.16,14	-0.37
TM4/3	0.36,8	0.02	0.39,11	0.04	0.08	0.10	0.51
TM7/3	-0.48	-0.02	-0.01	-0.23	-0.11	-0.08	-0.28
TM4/7	0.48,9	-0.01	0.11	0.11	0.19	0.11	0.48
TM5/3	-0.06	-0.02	-0.32	-0.14	-0.01	-0.11	-0.09
ND	0.35	-0.01	0.39	0.11	0.16	0.19	0.42
LOGTM3	-0.31	-0.13	-0.28	-0.02	-0.13	-0.10	0.08
LOGTM4	0.33	0.15	0.02	0.13	0.11	0.13	0.11
LOGTM5	-0.46	-0.17	-0.19	-0.03	-0.01	-0.02	-0.19
LOGTM7	-0.49	-0.02	-0.16	-0.02	-0.21	-0.15	-0.01

	W.wheat n=110	S.wheat n=26	Rye n=61	Barley n=108	Oats n=70	Potatoes n=46	Grazing crop, n=48
SSA2, JULY							
TM3	0.01	-0.27	-0.03	-0.11	-0.38	-0.12	-0.21
TM4	0.27,15	0.19	0.21,18	0.01	0.29,19	0.29,21	0.22
TM5	-0.30,16	-0.26	-0.16	-0.19	-0.11	-0.19	-0.21
TM7	-0.31	-0.41	-0.09	-0.12	-0.16	-0.12	-0.12
TM4/3	0.25	0.09	0.04	0.04	0.18	0.12	0.17
TM7/3	-0.36	-0.37	0.19	-0.13	-0.19	-0.13	-0.09
TM4/7	0.37,17	0.24	-0.15	0.17	0.24,20	0.21	0.11
TM5/3	-0.36	-0.14	0.03	-0.21	-0.19	-0.19	-0.21
ND	0.26	0.09	-0.19	0.21	0.19	0.23	0.19
LOGTM3	-0.01	-0.27	-0.09	-0.12	-0.38	-0.20	-0.15
LOGTM4	0.28	0.16	0.16	0.02	0.29	0.31	0.29
LOGTM5	-0.32	-0.34	-0.18	-0.12	-0.10	-0.12	-0.17
LOGTM7	-0.36	-0.25	-0.18	-0.11	-0.15	-0.09	-0.21

	W.Wheat n=41	S.Wheat n=15	Rye n=38	Barley n=35	Oats n=20	Potatoes n=62	Grazing crop, n=22
SSA3, JUNE							
TM3	-0.45	-0.55	-0.62	-0.14	-0.04	-0.67	-0.39
TM4	0.48,22	0.04	0.30	0.32	0.31	0.41	0.48
TM5	-0.40	-0.39	-0.61	-0.02	-0.15	-0.57	-0.38
TM7	-0.45	-0.14	-0.64	-0.10	-0.05	-0.64	-0.33
TM4/3	-0.41	0.49	0.48	0.27	0.13	0.56	0.17
TM7/3	0.39	-0.48	-0.55	-0.07	-0.06	-0.13	-0.46
TM4/7	-0.45	0.12	0.50	0.18	0.17	0.57	0.02
TM5/3	-0.31	0.55	-0.01	-0.25	-0.09	-0.42	-0.45
ND	0.32,23	0.36	0.56	0.27	0.21	0.61	0.44
LOGTM3	-0.44	-0.61	-0.62	-0.15	-0.06	-0.66	-0.42
LOGTM4	0.31	0.04	0.29	0.36	0.15	0.49	0.39
LOGTM5	-0.40	-0.41	-0.59	-0.01	-0.18	-0.53	-0.42
LOGTM7	-0.34	-0.15	-0.58	-0.07	-0.19	-0.59	-0.46

	W.Wheat n=41	S.Wheat n=15	Rye n=38	Barley n=35	Oats n=20	Potatoes n=62	Grazing crop, n=22
SSA3, JULY							
TM3	-0.24	-0.55	-0.23	-0.21	-0.28	-0.14	-0.48
TM4	0.15	0.34	0.21	0.31	0.32	0.09	0.36
TM5	-0.30	-0.39	-0.53	-0.18	-0.57	0.15	-0.17
TM7	-0.27	-0.14	-0.47	-0.19	-0.44	0.18	-0.35
TM4/3	0.37	0.49	0.11,24	0.28	0.04	0.03	0.05
TM7/3	-0.19	0.48	-0.45	0.17	-0.66	0.49	0.49
TM4/7	0.36	0.12	0.24	0.09	0.49	0.02	-0.02
TM5/3	0.02	0.55	-0.38	-0.01	-0.46	0.22	0.22
ND	0.37	0.43	0.14	0.34	-0.12	0.14	0.14
LOGTM3	-0.37	-0.62	-0.24	-0.20	-0.22	-0.10	-0.10
LOGTM4	0.11	0.29	0.24	0.29	0.34	0.12	0.12
LOGTM5	-0.37	-0.41	-0.53	-0.16	-0.53	0.11	0.11
LOGTM7	-0.38	-0.15	-0.45	-0.21	-0.43	0.17	0.17

	W.Wheat n=60	S.Wheat n=9	Rye n=53	Barley n=63	Oats n=33	Potatoes n=53	Grazing crop, n=33
SSA5, JUNE							
TM3	-0.50,25	-0.39	-0.40	-0.06	-0.26	-0.05	-0.25
TM4	0.17	0.21	0.07	0.09,30	-0.18	0.07	-0.25
TM5	-0.48	-0.53,28	-0.25	-0.08	-0.22	-0.09	-0.30
TM7	-0.49,26	-0.14	-0.32,29	-0.06	0.24	0.04	0.42,34
TM4/3	0.41	0.37	0.26	0.08	-0.10	-0.12	-0.25
TM7/3	-0.24	-0.04	-0.22	0.07	0.18	0.08	0.44
TM4/7	0.36	-0.34	0.09	-0.05	0.05	-0.03	0.23
TM5/3	0.27,27	-0.25	-0.06	0.09	0.21	0.09	-0.49
ND	0.39	0.26	0.30	0.05	-0.21	-0.09	-0.31
LOGTM4	-0.52	-0.31	-0.39	-0.01	0.11	0.11	0.45
LOGTM5	0.21	0.20	0.08	0.08	-0.16	-0.08	-0.19
LOGTM5	-0.39	-0.49	-0.21	-0.05	-0.21	-0.11	-0.27
LOGTM7	-0.37	-0.12	-0.30	-0.01	-0.12	-0.11	-0.10

	W.Wheat n=60	S.Wheat n=9	Rye n=53	Barley n=63	Oats n=33	Potatoes n=53	Grazing crop, n=33
SSA5, JULY							
TM3	-0.31	-0.29	-0.01	-0.42	-0.19	-0.32	-0.18
TM4	0.23	0.11	0.01	0.17	0.03	0.36,38	0.27
TM5	-0.57,35	-0.52	-0.42	-0.50	-0.28	-0.19	-0.16
TM7	-0.32	-0.21	-0.34	-0.43	-0.30	-0.35	-0.17
TM4/3	0.47,36	0.27	0.04	0.37	0.08	0.40	0.23
TM7/3	-0.11	-0.15	-0.35	-0.41,37	-0.26	-0.31	-0.05
TM4/7	0.35	0.46	0.23	0.48	0.24	0.39	0.13
TM5/3	-0.22	-0.04	-0.17	-0.07	-0.22	-0.05	-0.24
ND	0.35	0.31	0.02	0.39	0.12	0.35	0.26
LOGTM3	-0.22	-0.20	-0.01	-0.40	-0.21	-0.31	-0.11
LOGTM4	0.31	0.09	0.09	0.12	0.11	0.19	0.21
LOGTM5	-0.34	-0.42	-0.29	-0.45	-0.18	-0.34	-0.17
LOGTM7	-0.36	-0.11	-0.42	-0.39	-0.25	-0.31	-0.18

	W.Wheat n=31	S.Wheat n=10	Rye n=12	Barley n=34	Oats n=27	Grazing crop, n=25
SSA6, JUNE						
TM3	-0.32	-0.52	-0.23,40	-0.02	-0.11	-0.21
TM4	0.07	0.54	0.10	0.13	0.11	0.13
TM5	-0.41	-0.18	-0.03	-0.19	-0.27	-0.37,42
TM7	-0.47	-0.36	-0.17	-0.18,41	-0.35	-0.35
TM4/3	0.23,39	0.39	0.19	-0.10	0.02	0.09
TM7/3	-0.52	-0.21	-0.07	0.27	-0.01	-0.47
TM4/7	0.35	0.40	0.17	-0.20	0.02	0.23
TM5/3	0.03	-0.17	-0.12	0.12	-0.08	-0.43
ND	0.21	0.36	0.15	0.29	0.10	0.44
LOGTM4	-0.25	-0.46	-0.19	-0.01	-0.09	-0.12
LOGTM5	0.17	0.57	0.11	0.21	0.05	0.15
LOGTM5	-0.31	-0.19	-0.01	-0.15	-0.21	-0.41
LOGTM7	-0.41	-0.12	-0.02	-0.13	-0.19	-0.39

	W.Wheat n=31	S.Wheat n=10	Rye n=12	Barley n=34	Oats n=27	Grazing crop, n=25
SSA6, JULY						
TM3	-0.28	-0.20	-0.16,43	-0.38,44	-0.18	-0.30
TM4	0.31	0.22	0.23	0.02	0.15	0.11
TM5	-0.33	-0.38	-0.02	-0.41	-0.23	-0.23
TM7	-0.43	-0.05	-0.03	-0.37,45	-0.11	-0.36
TM4/3	0.29	0.57	0.21	0.43	0.12	0.31
TM7/3	-0.17	-0.39	-0.09	-0.14	-0.09	-0.13
TM4/7	0.21	0.11	0.39	0.35	-0.02	0.34
TM5/3	-0.21	-0.41	-0.03	-0.08	-0.14	-0.16,46
ND	0.11	0.46	0.31	0.36	0.12	0.29
LOGTM3	-0.21	-0.19	-0.15	-0.41	-0.16	-0.31
LOGTM4	0.01	0.21	0.17	0.13	0.19	0.12
LOGTM5	-0.31	-0.21	-0.01	-0.36	-0.16	-0.21
LOGTM7	-0.37	-0.01	-0.09	-0.29	-0.15	-0.25

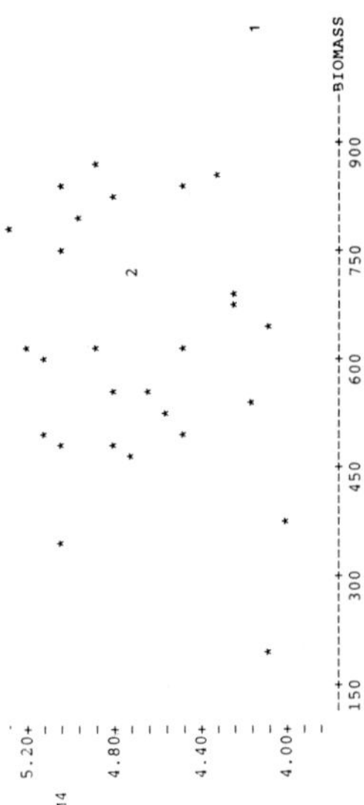
TM4
5.20+
4.80+
4.40+
4.00+
150
300
450
600
750
900
BIOMASS
1

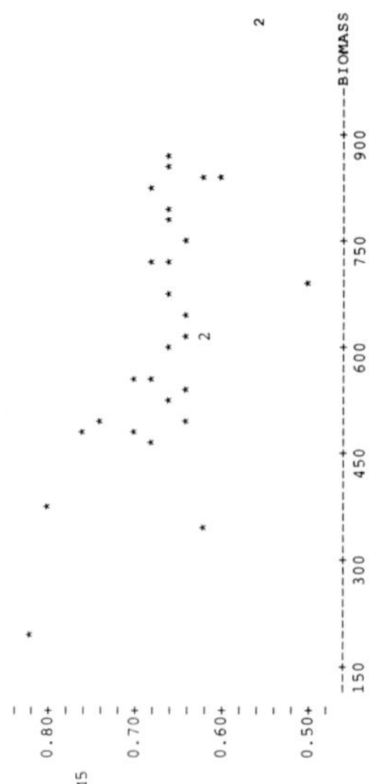
TM5
0.80+
0.70+
0.60+
0.50+
150
300
450
600
750
900
BIOMASS
2

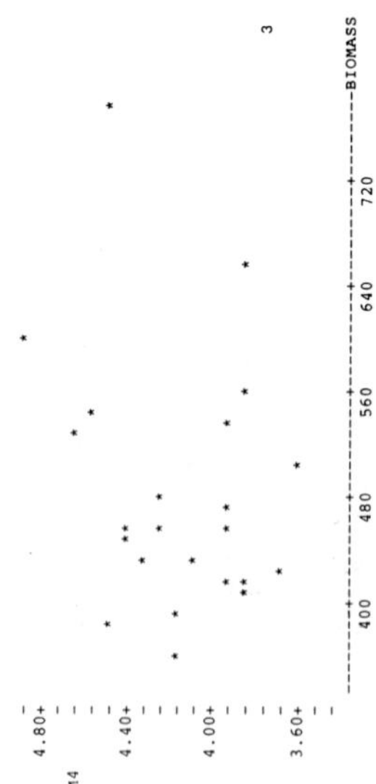
TM4
4.80+
4.40+
4.00+
3.60+
400
480
560
640
720
BIOMASS
3

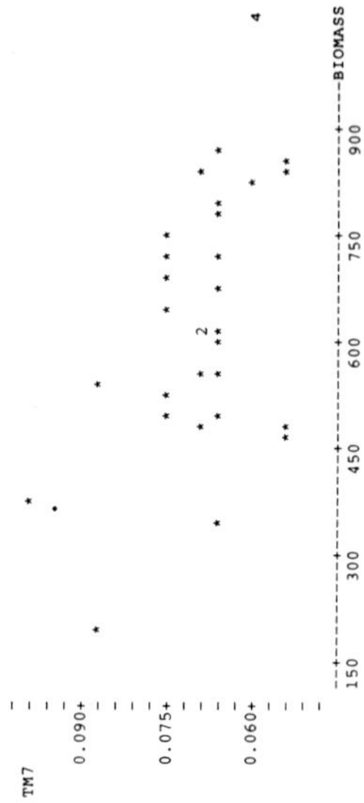
TM7
0.090+
0.075+
0.060+
150
300
450
600
750
900
BIOMASS
4

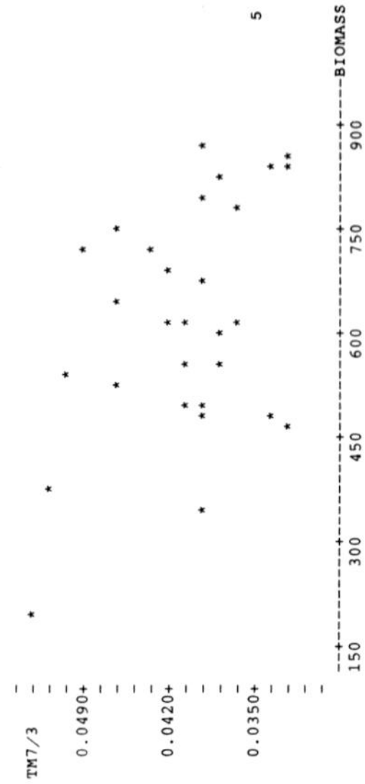
TM7/3
0.0490+
0.0420+
0.0350+
150
300
450
600
750
900
BIOMASS
5

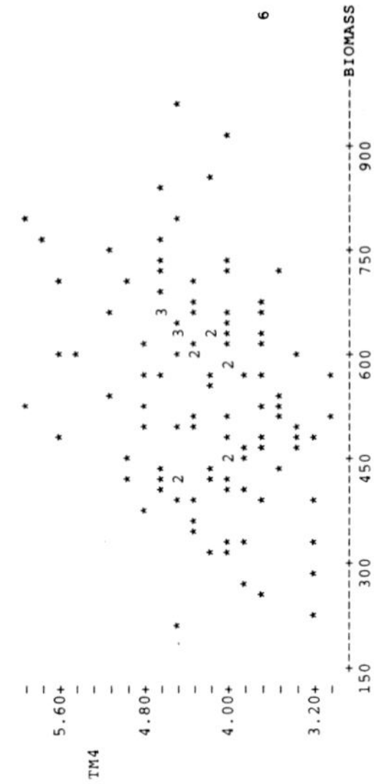
TM4
5.60+
4.80+
4.00+
3.20+
150
300
450
600
750
900
BIOMASS
6

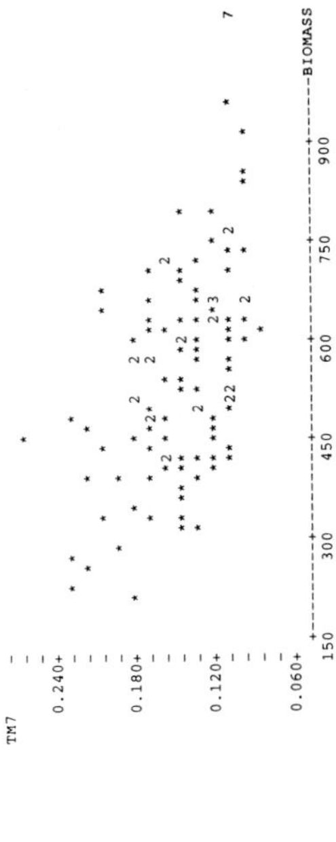
TM7
0.240+
0.180+
0.120+
0.060+
150
300
450
600
750
900
BIOMASS
7

TM4/3
4.0+
3.0+
2.0+
1.0+
150
300
450
600
750
900
BIOMASS
8

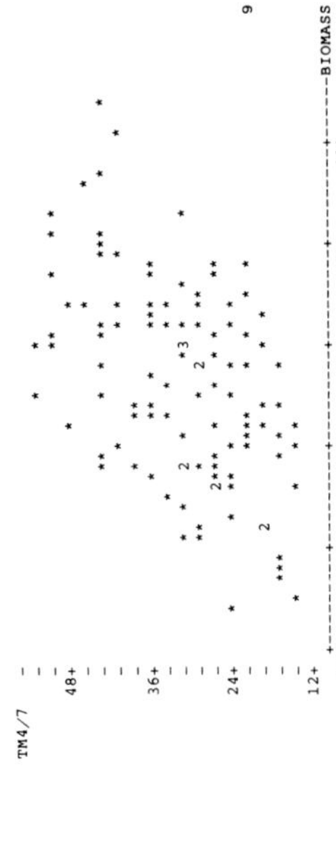
TM4/7
48+
36+
24+
12+
150
300
450
600
750
900
BIOMASS
9

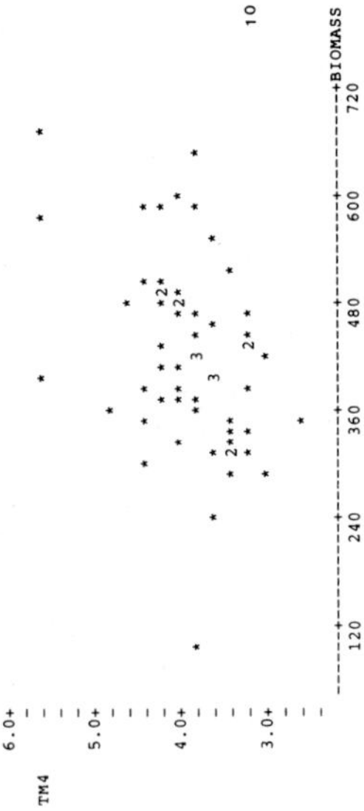
TM4
6.0+
5.0+
4.0+
3.0+
120
240
360
480
600
720
BIOMASS
10

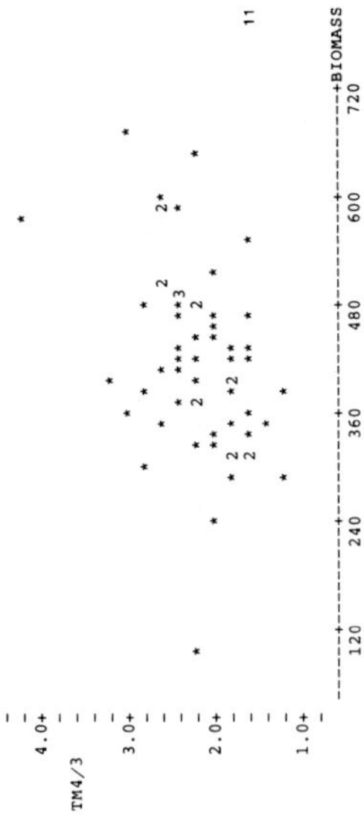
TM4/3
4.0+
3.0+
2.0+
1.0+
120
240
360
480
600
720
BIOMASS
11

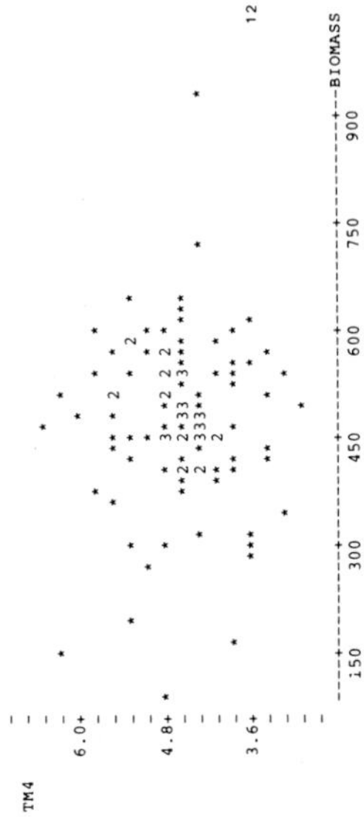
TM4
6.0+
4.8+
3.6+
150
300
450
600
750
900
BIOMASS
12

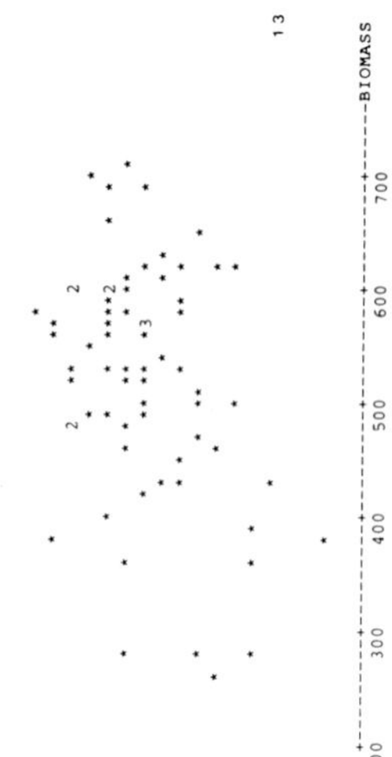

13

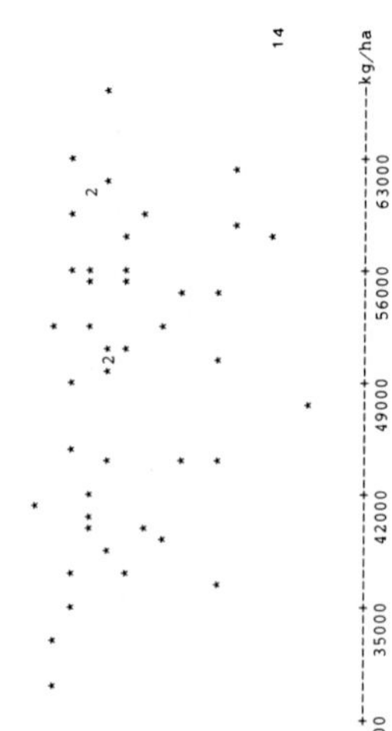

14

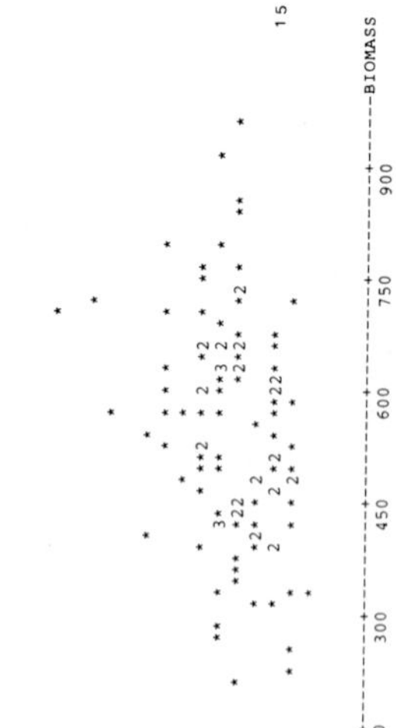

15

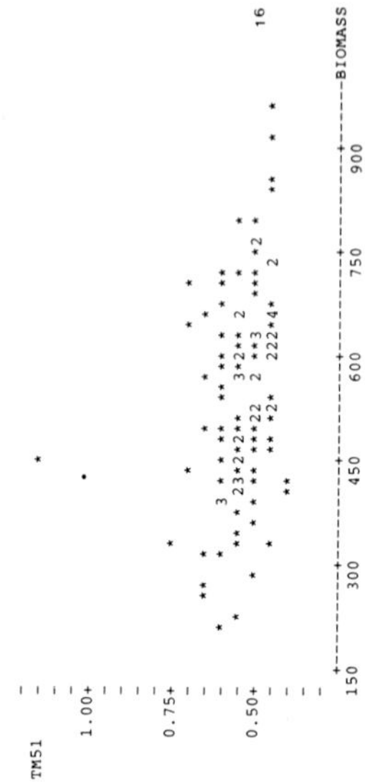

16

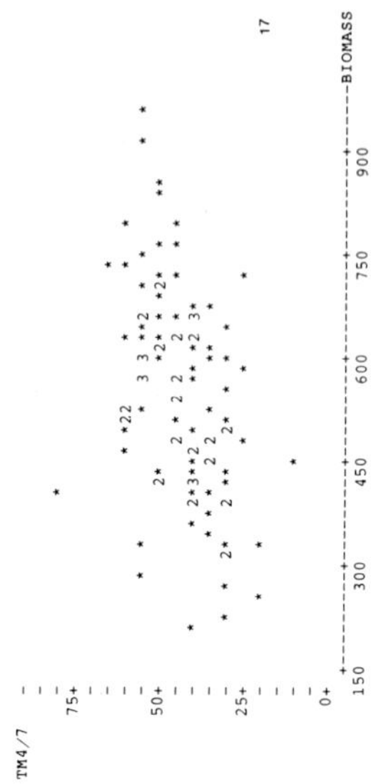

17

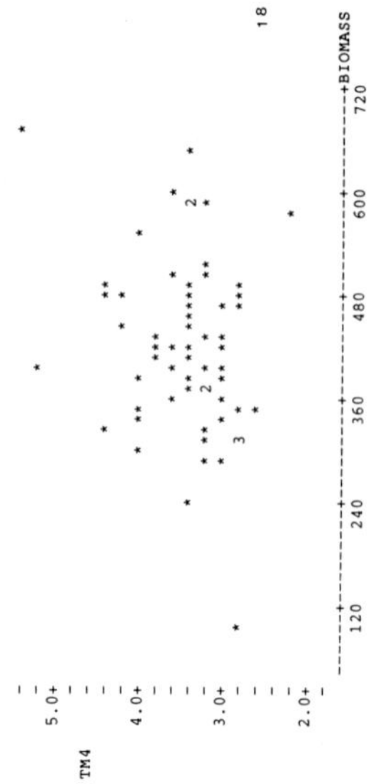

18

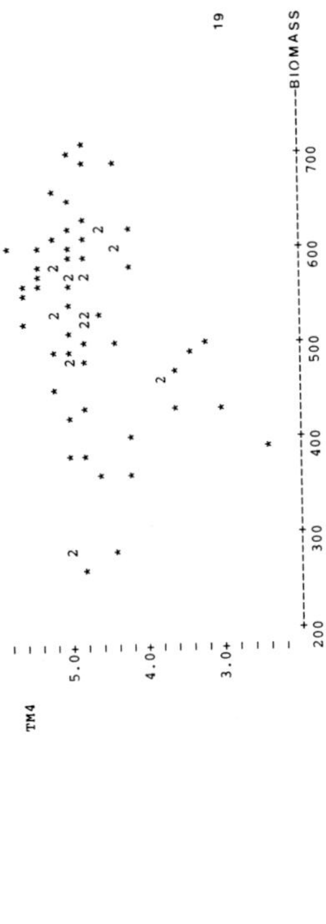
19
TM4
BIOMASS

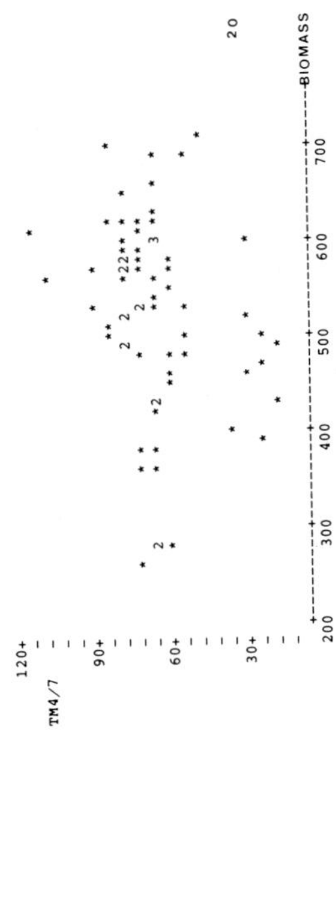
20
TM4/7
BIOMASS

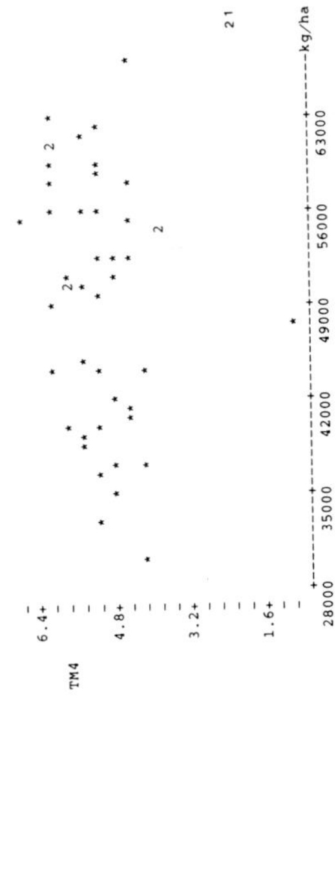
21
TM4
kg/ha

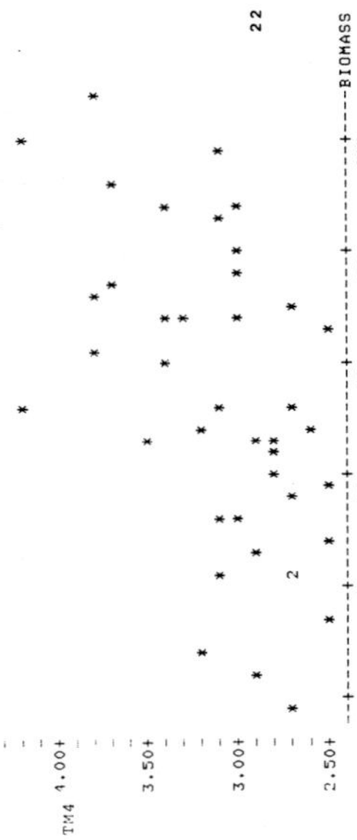
22
TM4
BIOMASS

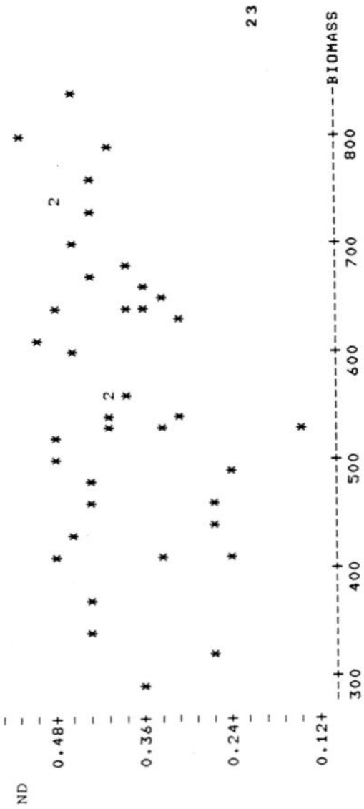
23
ND
BIOMASS

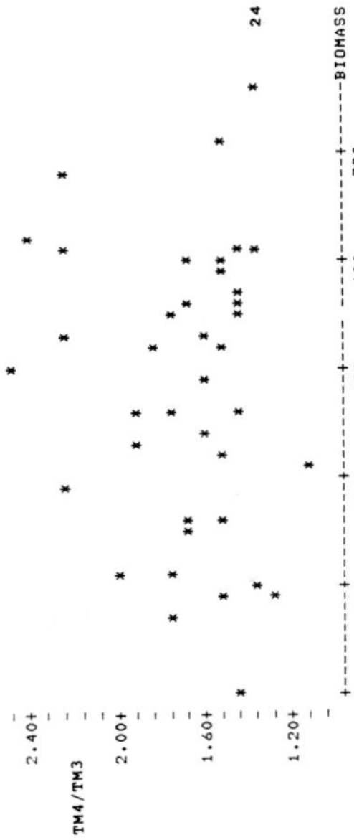
24
TM4/TM3
BIOMASS

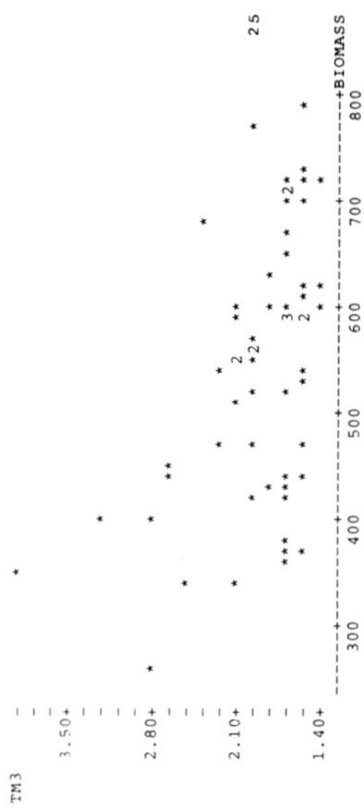
25
TM3
BIOMASS

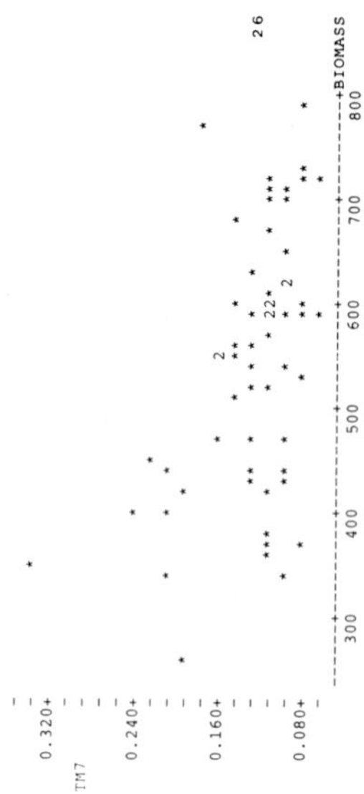
26
TM7
BIOMASS

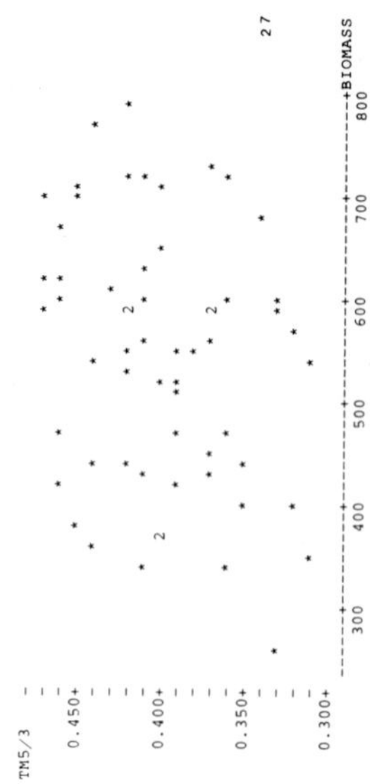
27
TM5/3
BIOMASS

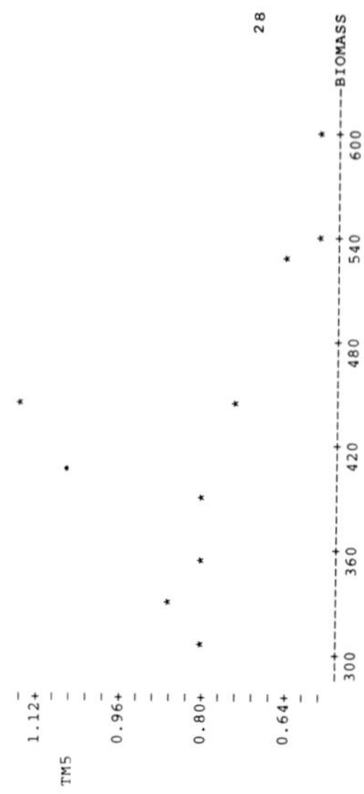
28
TM5
BIOMASS

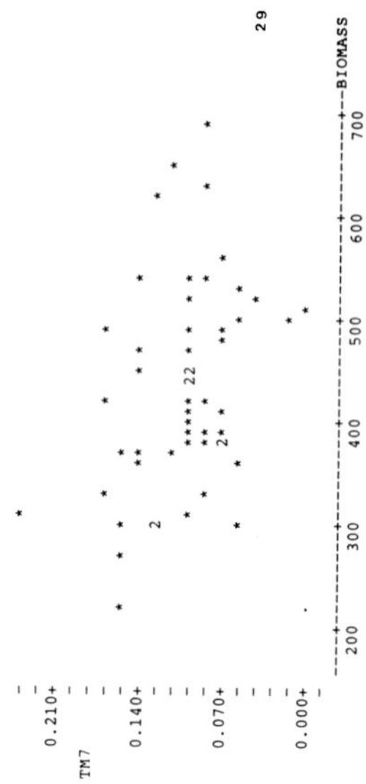
29
TM7
BIOMASS

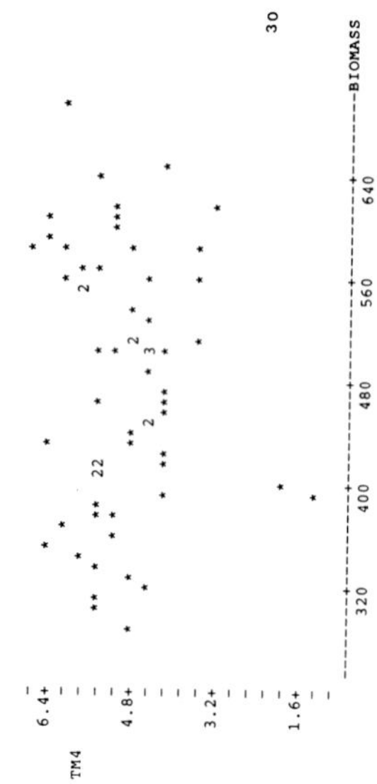
30
TM4
BIOMASS

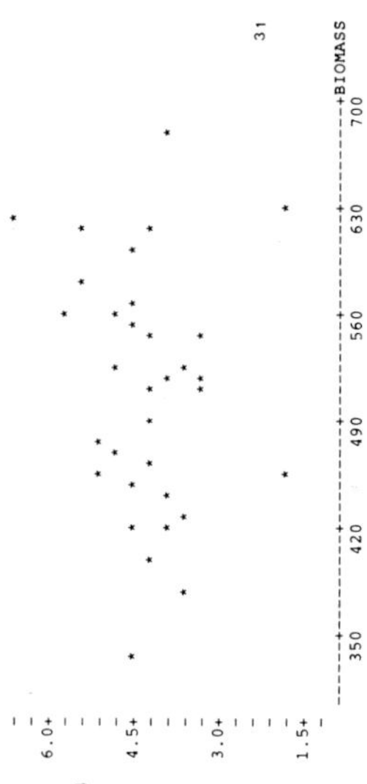
TM4
6.0+
4.5+
3.0+
1.5+
350
420
490
560
630
700
BIOMASS
31

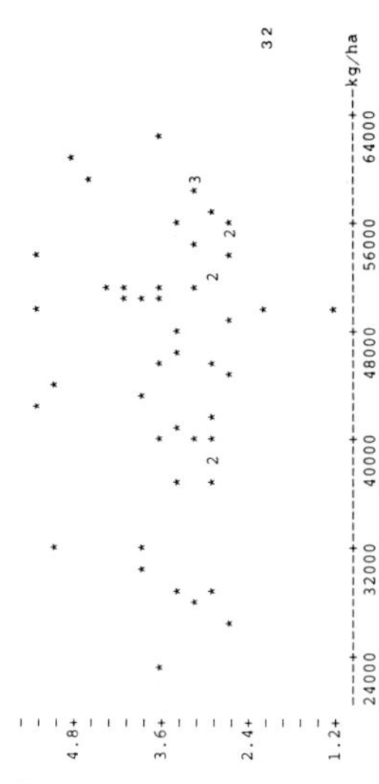
TM4
4.8+
3.6+
2.4+
1.2+
24000
32000
40000
48000
56000
64000
kg/ha
32

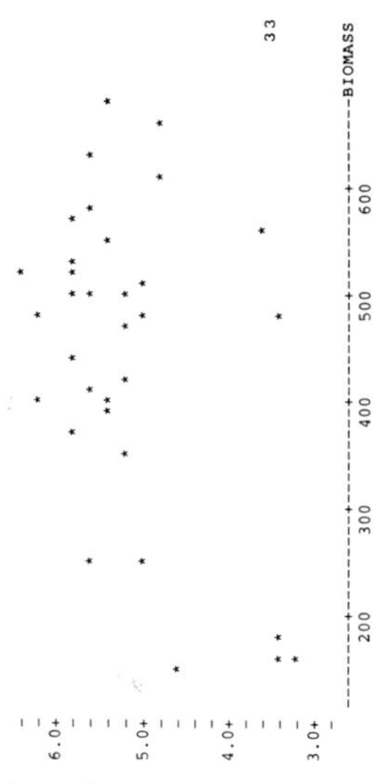
TM4
6.0+
5.0+
4.0+
3.0+
200
300
400
500
600
BIOMASS
33

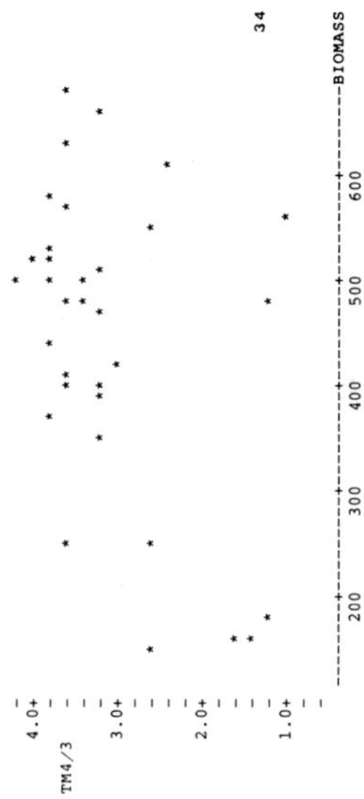
TM4/3
4.0+
3.0+
2.0+
1.0+
200
300
400
500
600
BIOMASS
34

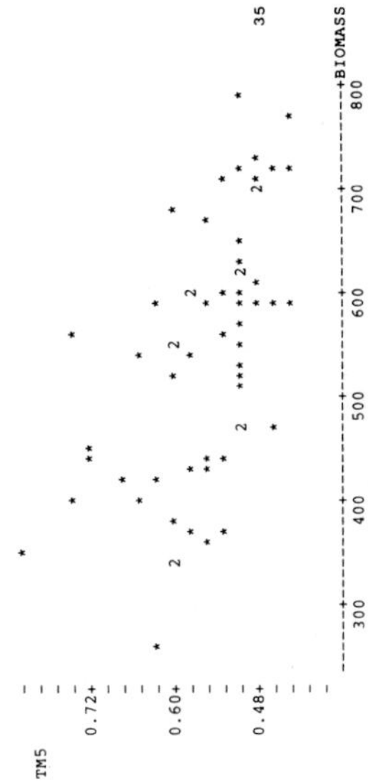
TM5
0.72+
0.60+
0.48+
300
400
500
600
700
800
BIOMASS
35

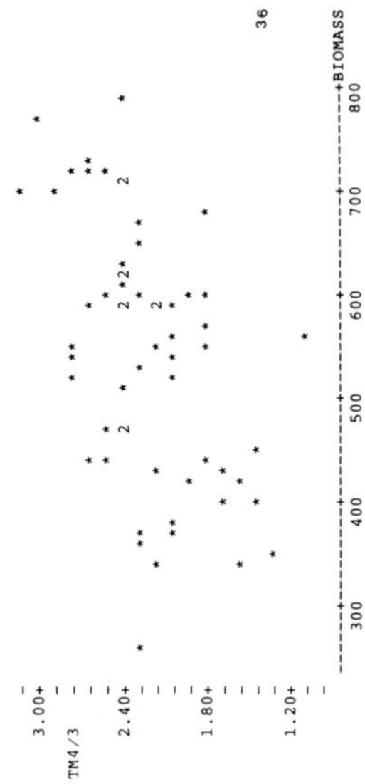
TM4/3
3.00+
2.40+
1.80+
1.20+
300
400
500
600
700
800
BIOMASS
36

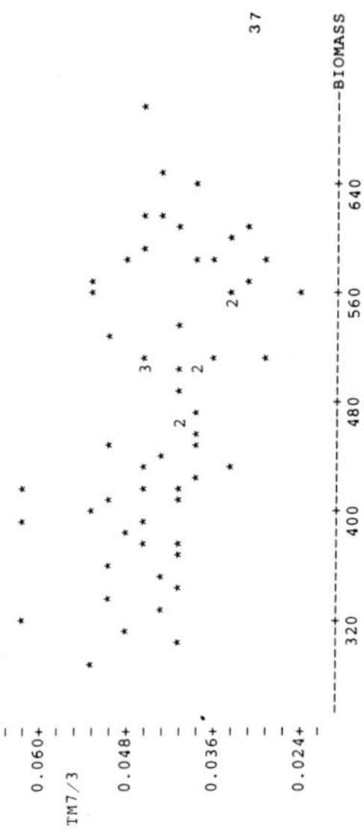
TM7/3
0.060+
0.048+
0.036+
0.024+
320
400
480
560
640
BIOMASS
37

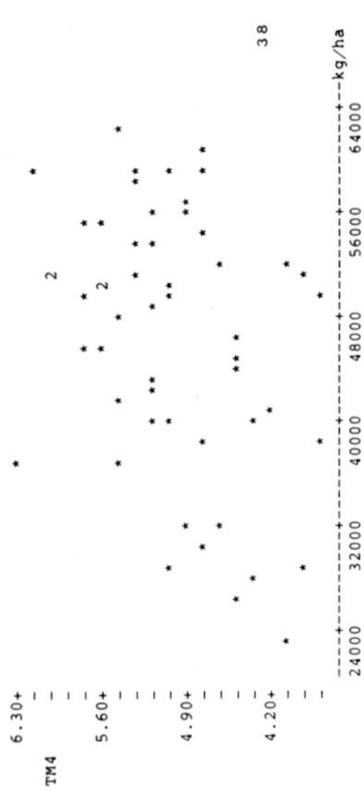
TM4
6.30+
5.60+
4.90+
4.20+
24000
32000
40000
48000
56000
64000
kg/ha
38

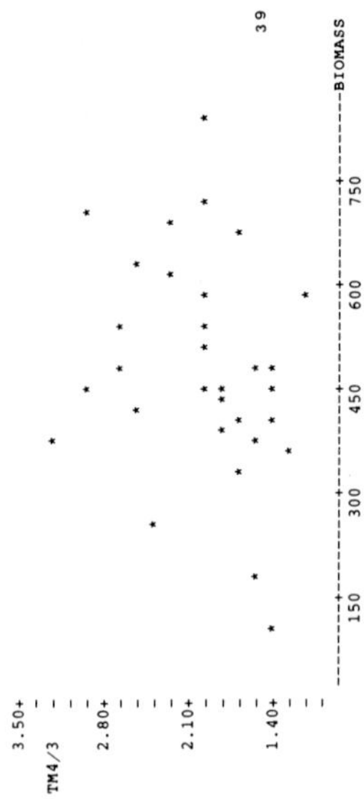
TM4/3
3.50+
2.80+
2.10+
1.40+
150
300
450
600
750
BIOMASS
39

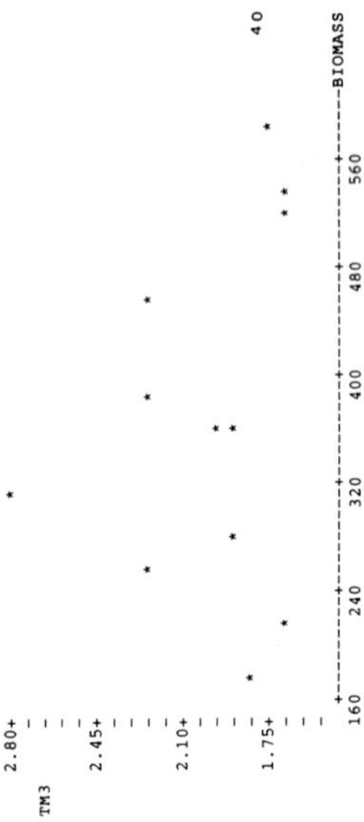
TM3
2.80+
2.45+
2.10+
1.75+
160
240
320
400
480
560
BIOMASS
40

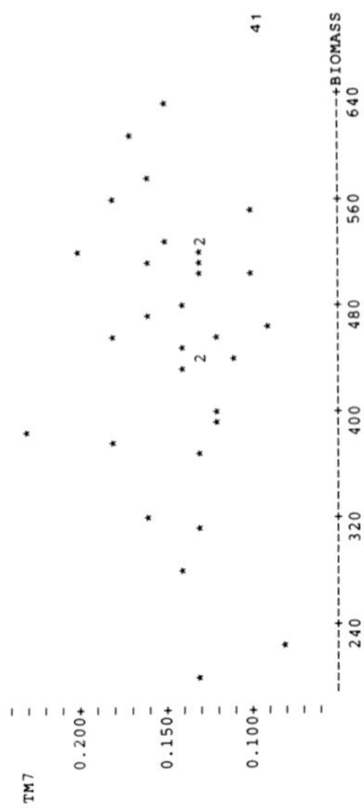
TM7
0.200+
0.150+
0.100+
240
320
400
480
560
640
BIOMASS
41

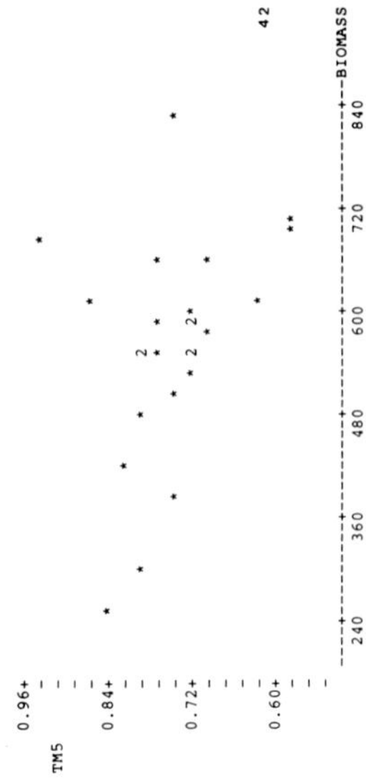
TM5
0.96+
0.84+
0.72+
0.60+
240
360
480
600
720
840
BIOMASS
42

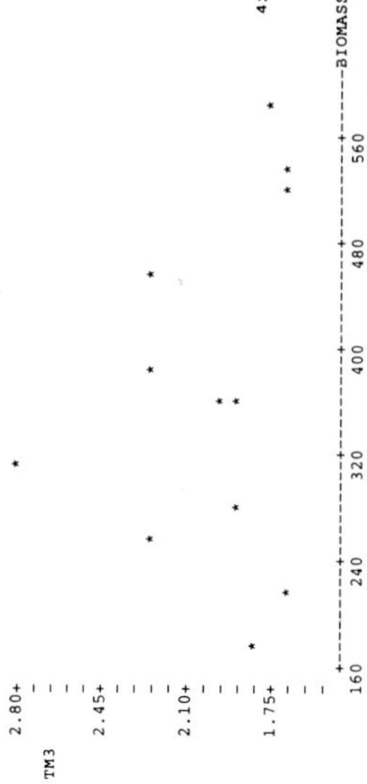
TM3
2.80+
2.45+
2.10+
1.75+
160
240
320
400
480
560
BIOMASS

43

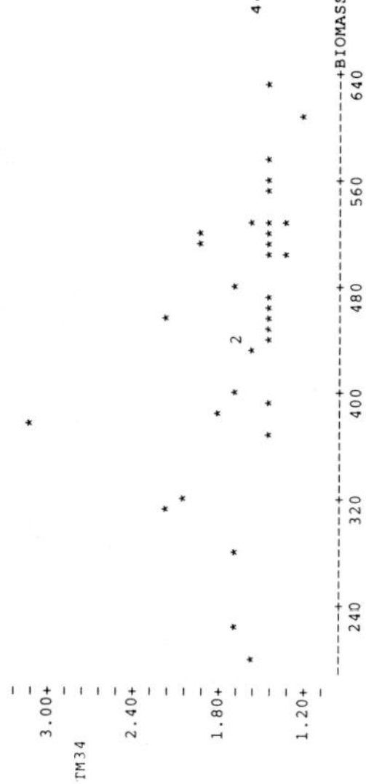
TM34
3.00+
2.40+
1.80+
1.20+
240
320
400
480
560
640
BIOMASS

44

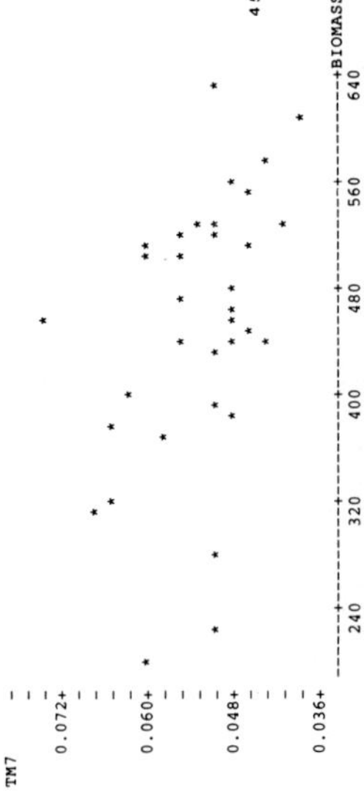
TM7
0.072+
0.060+
0.048+
0.036+
240
320
400
480
560
640
BIOMASS

45

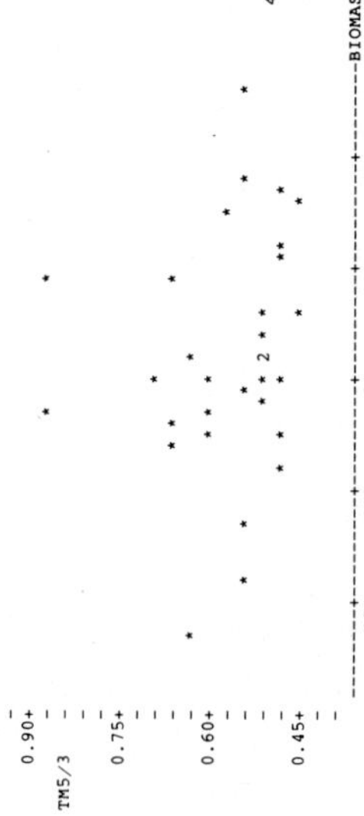
TM5/3
0.90+
0.75+
0.60+
0.45+
150
300
450
600
750
BIOMASS

46

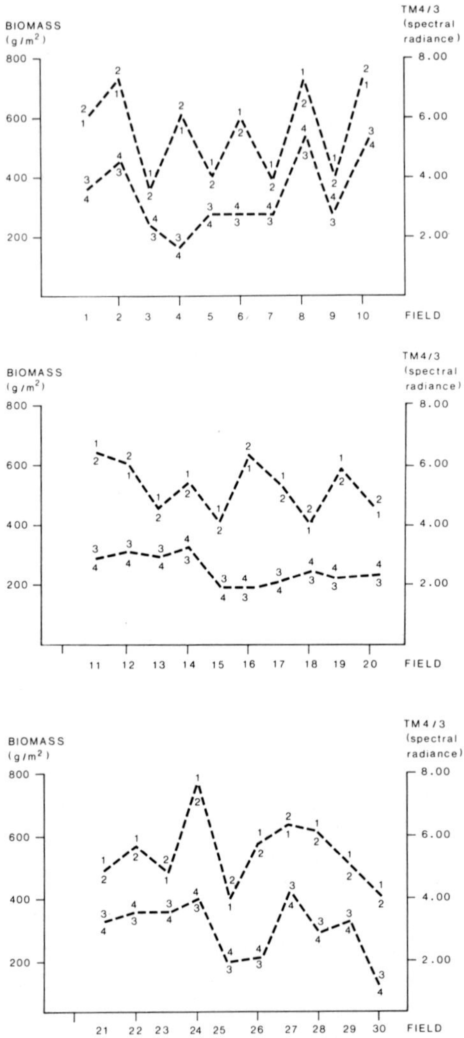

The biomass (g/m²) and TM data of the two plots within each field.
1= the biomass of plot one
2= the biomass of plot two
3= TM data of plot one
4= TM data of plot two

APPENDIX D

THE KAPPA COEFFICIENT OF AGREEMENT AS A MEASURE OF TOTAL MAP ACCURACY AND OVERALL ACCURACY

1-15= transformations 1-15 (Tab. 7.1).

TRANS-FORM	KAPPA			OVERALL		
	MEAN	MIN	MAX	MEAN	MIN	MAX
1.						
S.A. 1	53.1	24.0	82.2	69.5	56.2	82.8
S.A. 2	59.5	50.1	68.9	65.8	61.3	70.3
S.A. 3	61.0	45.8	76.2	67.2	60.1	74.3
S.A. 4	44.3	23.9	64.6	54.8	44.7	64.9
S.A. 5	60.3	48.7	71.9	66.7	58.5	74.8
S.A. 6	58.5	41.6	75.4	57.7	52.1	63.3
2.						
S.A. 1	61.0	32.1	89.9	76.1	63.7	88.4
S.A. 2	56.9	47.8	66.5	63.4	58.9	67.9
S.A. 3	60.9	45.7	76.0	67.2	60.1	74.3
S.A. 4	8.1	0.0	28.5	56.9	46.9	67.1
S.A. 5	55.0	43.5	66.7	62.1	53.9	70.2
S.A. 6	50.1	33.3	67.0	57.7	52.1	63.3
3.						
S.A. 1	55.4	26.0	84.8	71.7	58.7	84.7
S.A. 2	53.5	44.1	62.9	60.3	55.8	64.8
S.A. 3	55.2	40.1	70.4	62.5	55.1	69.8
S.A. 4	47.7	27.3	68.1	56.9	46.9	67.1
S.A. 5	52.0	40.3	63.6	59.3	51.1	67.4
S.A. 6	50.0	33.2	66.8	57.7	52.1	63.3
4.						
S.A. 1	51.4	21.8	81.0	67.3	53.8	80.9
S.A. 2	51.2	41.9	60.6	58.2	53.7	62.7
S.A. 3	53.1	38.0	68.2	60.7	53.3	68.1
S.A. 4	44.5	24.2	64.8	53.7	43.6	63.8
S.A. 5	51.9	40.3	63.5	59.4	51.2	67.5
S.A. 6	51.7	34.9	68.6	59.2	53.5	64.8
5.						
S.A. 1	58.9	29.2	88.6	73.9	61.2	86.6
S.A. 2	55.0	45.7	64.3	61.6	57.1	66.1
S.A. 3	53.0	37.9	68.2	65.4	58.2	72.6
S.A. 4	44.1	23.8	64.4	52.6	42.5	62.8
S.A. 5	53.9	42.3	65.5	61.8	53.6	69.9
S.A. 6	50.1	33.2	66.9	57.7	52.1	63.3
6.						
S.A. 2	51.1	41.8	60.4	58.9	54.4	63.4
7.						
S.A. 1	44.0	14.9	73.1	63.0	49.0	76.9
S.A. 2	59.3	49.9	68.7	65.3	60.8	69.8
S.A. 3	48.7	33.6	63.9	55.9	48.4	63.4
S.A. 4	56.8	36.4	77.2	64.5	54.8	74.2
S.A. 5	53.7	42.1	65.3	60.4	52.2	68.5
S.A. 6	59.3	42.5	76.2	66.6	60.9	72.2
8.						
S.A. 1	45.2	15.7	75.3	65.2	51.4	78.9
S.A. 2	59.6	50.3	69.0	65.7	61.3	70.2
S.A. 3	51.0	35.9	66.1	56.5	49.0	64.1
S.A. 4	46.2	25.9	66.5	55.9	45.8	66.0
S.A. 5	53.9	42.3	65.6	60.4	52.2	68.5
S.A. 6	56.8	39.9	73.7	64.4	58.7	70.1
9.						
S.A. 1	34.3	47.3	63.9	56.5	42.2	70.8
S.A. 2	56.0	46.7	65.4	65.7	61.3	70.1
S.A. 3	44.8	29.6	59.9	52.3	44.8	59.9
S.A. 4	41.2	20.8	61.6	52.6	42.5	62.8
S.A. 5	48.2	36.5	59.9	55.5	47.3	63.6
S.A. 6	56.6	39.8	73.4	63.7	58.0	69.3
10.						
S.A. 1	83.3	54.3	100.0	91.3	77.4	100.0
S.A. 2	52.3	42.9	61.7	59.8	55.2	64.4
S.A. 3	42.6	27.5	57.7	51.1	43.6	58.7
S.A. 4	49.8	29.4	70.1	60.2	50.2	70.1
S.A. 5	49.3	60.9	37.7	56.9	48.7	65.0
S.A. 6	38.8	22.0	55.7	48.8	43.1	54.5
11.						
S.A. 2	54.9	45.6	64.3	62.3	57.8	66.8
12.						
S.A. 2	54.5	45.1	63.9	61.8	57.3	66.4
13.						
S.A. 2	59.5	50.1	68.9	65.7	61.3	70.1
14.						
S.A. 2	64.4	55.0	73.8	70.1	65.8	74.3
15.						
S.A. 2	57.4	48.1	66.8	63.9	59.4	68.4

MIN= minimum 95% confidence interval
MAX= maximum 95% confidence interval
S.A. = sub-area

APPENDIX E

CONDITIONAL KAPPA, MEAN, PRODUCER'S AND USER'S ACCURACY

TRANS-FORM 3	MEAN	MIN	MAX	MEAN	MIN	MAX	MEAN	MIN	MAX	MEAN	MIN	MAX
	KAPPA			MEAN ACC.			PRODUCER'S			USER'S		
S.A. 1												
W.wheat	29.6	1.4	57.8	64.8	50.4	79.3	100.0			48.0	33.5	62.1
Rye	91.2	62.3	100.0	95.2	88.8	100.0	95.2	89.9	100.0	95.2	89.9	100.0
S.A. 2												
W.wheat	40.1	30.8	49.5	59.5	54.9	64.1	73.6	69.4	77.3	50.0	45.1	54.6
S.wheat	35.9	45.3	54.8	68.0	63.8	72.5	74.4	69.9	78.1	63.0	58.5	67.5
Rye	74.4	65.0	83.8	67.2	62.0	72.5	52.9	48.2	57.6	92.3	89.1	94.8
Barley	63.0	53.6	72.3	68.5	64.1	72.8	82.2	78.6	85.8	58.7	54.1	63.3
Oats	60.5	51.2	69.9	58.7	54.1	63.3	70.2	65.9	74.5	50.4	45.8	55.1
Potatoes	88.2	78.8	97.7	69.1	64.8	73.4	60.5	55.9	65.1	80.7	77.0	84.3
Gr.crop	59.4	50.0	68.8	37.2	32.7	41.8	31.4	27.0	35.7	45.8	41.1	50.5
S.A. 3												
W.wheat	51.3	36.2	66.4	67.7	60.7	74.7	78.5	72.3	84.7	59.4	52.0	66.8
S.wheat	53.9	38.6	69.2	83.1	77.5	88.7	78.0	71.7	84.3	88.8	84.1	93.6
Rye	81.3	66.1	96.4	45.4	37.9	52.9	83.3	77.6	88.5	31.2	24.2	38.2
Barley	59.4	44.3	74.5	94.1	90.6	97.6	74.1	67.5	80.8	80.0	73.9	86.0
Oats	39.4	22.4	54.4	57.1	49.6	64.6	73.3	67.4	80.7	50.0	42.4	57.5
Potatoes	48.6	33.6	63.7	32.0	25.0	39.0	28.5	12.6	24.3	26.6	19.9	33.3
Gr.crop	82.8	67.1	97.9	66.6	59.5	73.7	78.9	72.7	85.1	52.9	45.3	60.4
S.A. 4												
W.wheat	37.1	17.1	57.2	58.5	48.5	68.5	75.0	66.1	83.8	48.0	37.8	58.1
Rye	83.1	62.8	100.0	57.1	47.0	67.2	39.8	50.0	60.1	66.6	57.0	76.2
Barley	11.4	0.0	31.4	35.2	25.5	45.0	23.1	14.5	31.6	75.0	66.1	83.8
Oats	68.5	47.4	89.7	57.7	47.7	67.8	86.7	79.7	93.5	43.3	33.2	53.4
Potatoes	65.1	45.8	84.5	82.9	75.2	90.5	60.0	50.0	70.0	25.0	16.1	33.8
S.A. 5												
W. wheat	48.8	37.2	60.4	65.9	60.0	71.8	81.0	75.1	86.9	55.5	49.6	61.4
Rye	87.0	75.3	98.7	78.3	72.4	89.6	69.2	63.3	75.1	90.0	84.1	95.9
Barley	55.0	43.4	66.7	66.6	60.7	72.5	69.6	64.0	75.8	63.9	58.0	69.8
Oats	53.0	41.4	64.7	59.0	53.1	64.9	60.0	54.1	65.9	58.0	52.1	63.9
Potatoes	46.1	34.5	57.7	65.9	60.0	61.8	88.2	82.3	94.1	52.6	46.7	57.9
Gr.cr.	87.9	76.2	99.6	68.3	62.4	74.2	55.1	49.2	61.0	90.0	84.1	96.0
S.A. 6												
W.wheat	40.7	23.9	57.5	55.5	46.9	64.1	60.0	51.4	68.6	51.7	43.1	60.3
S.wheat	20.8	3.9	37.7	27.2	18.6	35.8	27.2	18.6	35.8	27.2	18.6	35.8
Rye	89.7	72.5	100.0	59.4	50.8	68.0	44.4	35.8	53.0	91.6	83.0	100.0
Barley	56.0	39.2	72.0	70.1	61.5	78.7	76.9	68.3	85.5	64.5	55.9	73.1
Oats	38.5	22.0	55.0	59.4	50.8	68.0	91.6	83.0	100.0	44.4	35.8	53.0
Gr.crop	95.0	78.0	100.0	83.3	74.7	91.9	74.2	65.6	82.8	96.2	87.6	100.0

MIN= minimum 95% confidence interval
MAX= maximum 95% confidence interval
S.A.= sub-area

TRANS-FORM 4	MEAN	MIN	MAX	MEAN	MIN	MAX	MEAN	MIN	MAX	MEAN	MIN	MAX
	KAPPA			MEAN ACC.			PRODUCER'S			USER'S		
S.A. 1												
W.wheat	40.6	12.1	69.1	95.2	88.7	100.0	100.0			60.0	45.8	74.1
Rye	91.2	62.3	100.0				95.2	89.0	100.0	95.2	89.0	100.1
S.A. 2												
W.wheat	43.9	34.5	53.2	39.3	34.7	43.9	32.4	28.0	36.8	50.0	45.3	54.6
S.wheat	41.5	32.1	50.9	62.3	57.7	66.8	74.0	69.9	78.1	53.7	49.1	58.4
Rye	76.5	67.1	85.9	67.4	63.0	71.8	52.9	48.2	57.6	63.0	58.5	67.5
Barley	49.0	39.7	58.4	62.1	57.5	66.6	82.2	78.6	85.8	92.3	89.8	94.8
Oats	57.3	47.9	66.6	63.1	58.6	67.6	70.2	65.9	74.5	57.1	52.5	61.7
Potatoes	90.8	81.4	100.0	63.4	58.9	67.9	60.5	55.9	65.1	51.4	46.7	56.1
Gr.crop	59.3	49.9	68.7	68.6	64.2	72.9	31.4	27.0	35.7	82.4	78.8	86.0
S.A. 3												
W.wheat	52.9	37.8	68.0	50.0	42.5	57.5	43.7	36.2	51.2	58.3	50.8	65.7
S.wheat	53.6	38.3	68.9	68.9	61.9	75.2	71.4	64.5	78.2	66.6	59.5	73.8
Rye	85.3	70.1	100.0	48.6	41.0	56.2	45.0	37.4	52.5	52.9	45.3	60.4
Barley	60.0	44.8	75.1	67.6	60.5	74.7	63.6	56.7	70.5	62.1	54.8	69.5
Oats	40.5	25.8	55.2	63.4	56.1	70.7	74.0	67.4	80.7	55.5	48.0	63.0
Potatoes	42.2	27.2	57.2	25.2	20.8	29.6	18.5	12.6	24.3	31.2	24.2	38.2
Gr.crop	76.8	61.6	92.0	76.9	70.5	83.3	78.9	72.7	85.1	75.0	68.4	81.5
S.A. 4												
W.wheat	43.4	23.5	63.4	87.5	77.3	97.7	87.5	77.6	97.4	87.5	79.0	96.0
Rye	83.1	62.8	100.0	63.1	53.3	72.9	92.3	82.9	100.0	64.8	56.3	73.3
Barley	15.1	0.0	35.4	53.3	43.1	63.4	44.4	34.3	54.5	66.6	58.1	75.1
Oats	69.0	47.8	90.1	57.1	47.1	67.2	46.1	36.0	56.2	75.0	66.5	83.5
Potatoes	63.5	42.7	84.3	46.1	36.0	56.2	54.5	44.2	64.6	72.0	63.5	80.5
S.A. 5												
W.wheat	39.4	27.8	51.0	58.1	52.1	63.9	78.1	72.2	84.0	46.2	40.2	52.0
Rye	94.6	82.9	1.00	79.3	73.4	85.2	67.6	61.6	73.4	96.0	90.1	100.0
Barley	31.6	20.0	43.1	51.6	45.7	57.5	75.0	69.1	81.0	39.3	33.4	45.2
Oats	56.5	44.8	68.1	61.3	55.4	67.2	61.2	55.4	67.2	61.2	55.3	67.1
Potatoes	47.7	36.0	59.3	58.9	53.0	69.8	60.0	54.1	65.9	57.8	51.9	63.7
Gr.crop	87.9	76.1	99.6	72.0	66.1	77.9	60.0	54.1	65.9	90.0	84.1	95.9
S.A. 6												
W.wheat	24.4	7.7	41.1	42.5	33.9	51.1	55.5	46.9	64.1	34.4	25.8	43.0
S.wheat	18.8	1.8	35.9	24.0	15.4	32.6	21.4	12.8	30.0	27.2	18.6	35.8
Rye	82.0	65.3	98.7	55.8	47.2	64.4	38.7	30.1	47.3	100.0		
Barley	46.9	30.2	63.6	66.6	58.0	75.2	85.0	76.4	93.6	54.8	46.2	63.4
Oats	34.6	18.2	51.0	55.5	46.9	64.1	90.9	82.3	99.5	40.0	31.4	48.6
Gr.crop	95.1	78.2	100.0	89.6	81.0	98.2	83.8	75.2	92.4	96.3	87.7	100.0

MIN= minimum 95% confidence interval
MAX= maximum 95% confidence interval
S.A.= sub-area

APPENDIX F

INDIVIDUAL CROP CLASSIFICATION ACCURACY

Trans.= transformation (Tab. 7.1)
S.A.= sub-area (Fig. 4.1)
1= winter wheat
2= summer wheat
3= rye
4= barley
5= oats
6= potatoes
7= grazing crop

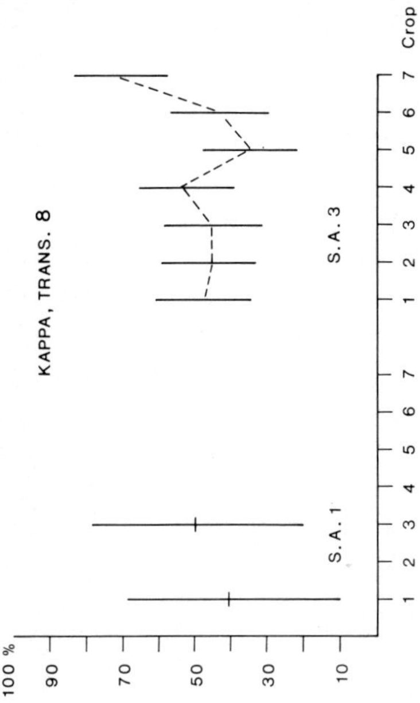
KAPPA, TRANS. 8
S.A. 1
S.A. 3
100 %
90
70
50
30
10
1
2
3
4
5
6
7
Crop

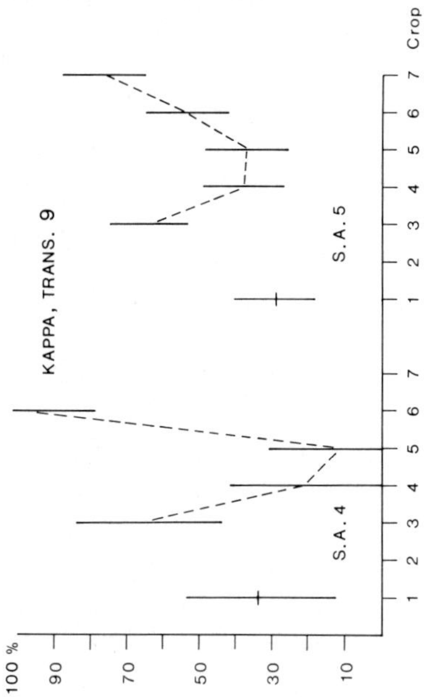
KAPPA, TRANS. 9
S.A. 4
S.A. 5
100 %
90
70
50
30
10
1
2
3
4
5
6
7
Crop

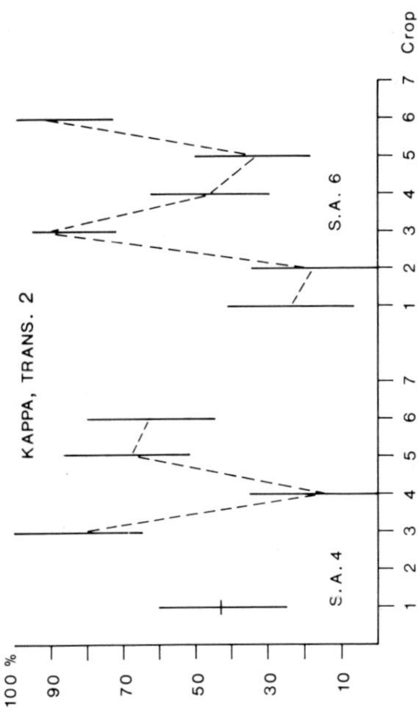
KAPPA, TRANS. 2
S.A. 4
S.A. 6
100 %
90
70
50
30
10
1
2
3
4
5
6
7
Crop

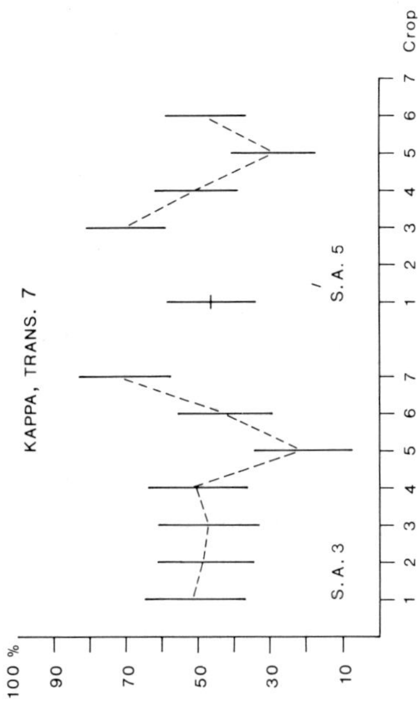
KAPPA, TRANS. 7
S.A. 3
S.A. 5
100 %
90
70
50
30
10
1
2
3
4
5
6
7
Crop

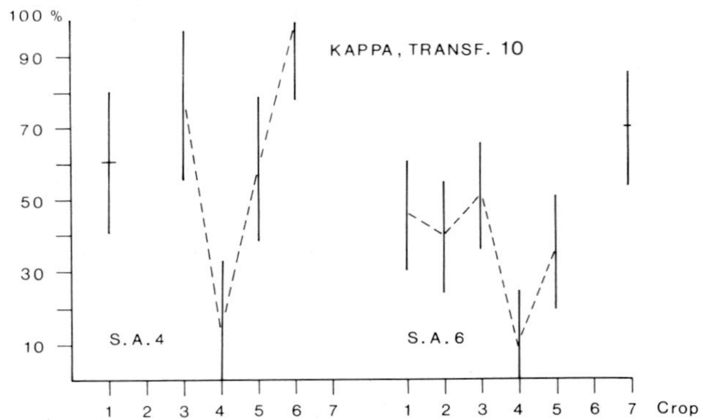
KAPPA, TRANSF. 10
100 %
90
70
50
30
10
S.A.4
S.A.6
1 2 3 4 5 6 7 1 2 3 4 5 6 7 Crop

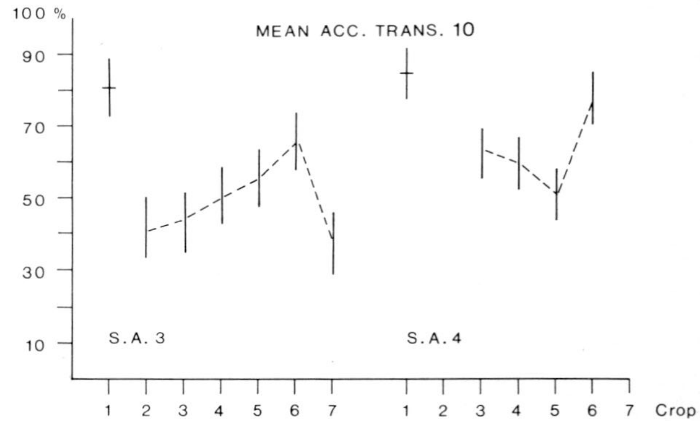
MEAN ACC. TRANS. 10
100 %
90
70
50
30
10
S.A.3
S.A.4
1 2 3 4 5 6 7 1 2 3 4 5 6 7 Crop

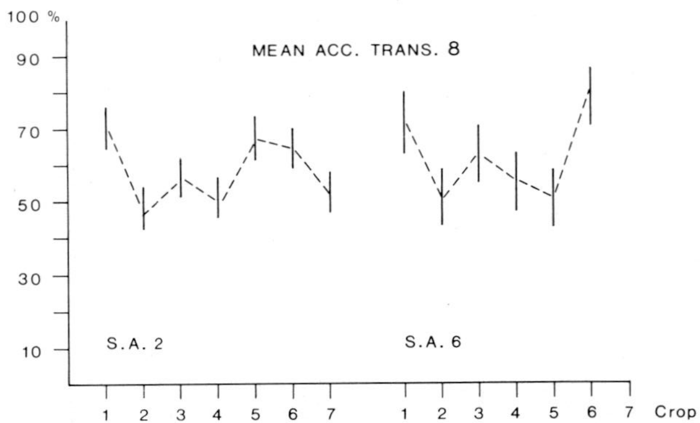
MEAN ACC. TRANS. 8
100 %
90
70
50
30
10
S.A.2
S.A.6
1 2 3 4 5 6 7 1 2 3 4 5 6 7 Crop

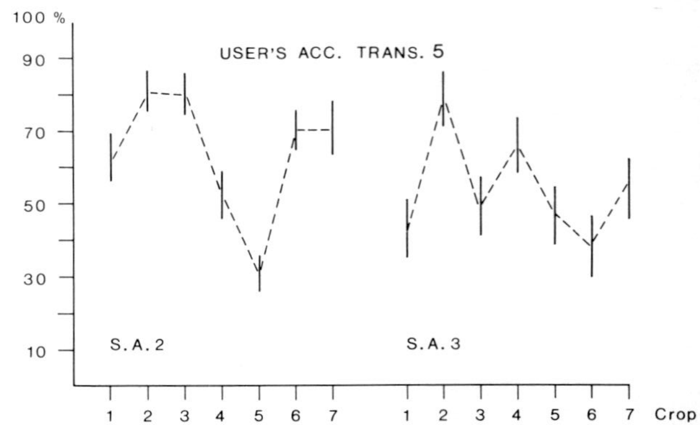
USER'S ACC. TRANS. 5
100 %
90
70
50
30
10
S.A.2
S.A.3
1 2 3 4 5 6 7 1 2 3 4 5 6 7 Crop

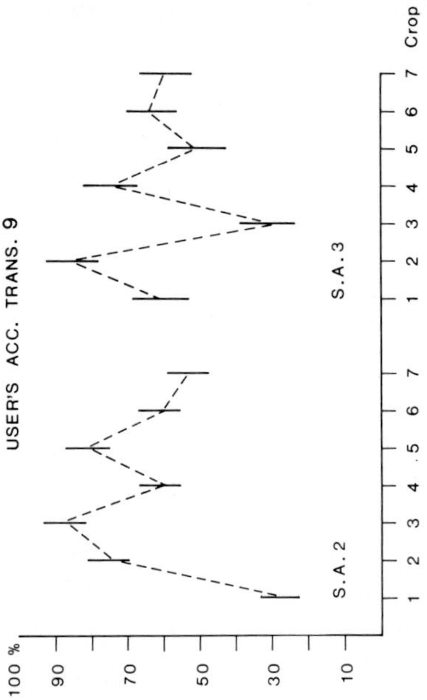
USER'S ACC. TRANS. 9
100 %
90
70
50
30
10
S.A.2
S.A.3
1 2 3 4 5 6 7
1 2 3 4 5 6 7
Crop

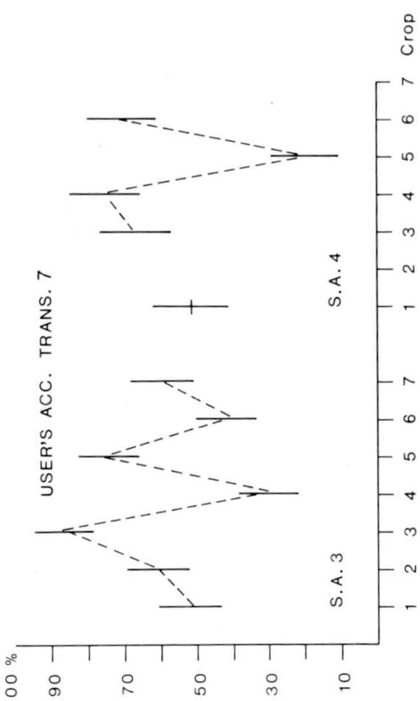
USER'S ACC. TRANS. 7
100 %
90
70
50
30
10
S.A.3
S.A.4
1 2 3 4 5 6 7
1 2 3 4 5 6 7
Crop

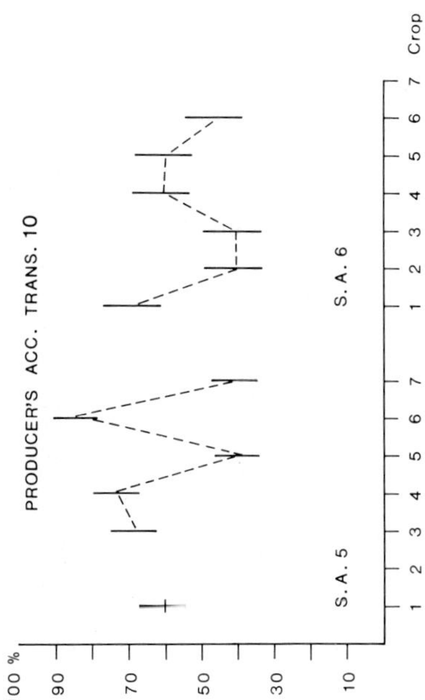
PRODUCER'S ACC. TRANS. 10
100 %
90
70
50
30
10
S.A.5
S.A.6
1 2 3 4 5 6 7
1 2 3 4 5 6 7
Crop

APPENDIX G

SUMMATION OF INDIVIDUAL CROP CLASSIFICATION RESULTS

	KAPPA ACC.	PRODUCER'S ACC.	MEAN ACC.	USER'S ACC.
WINTER WHEAT				
Highest acc.:	tr. 11-15, appr. 50%	tr. 4, appr. 100%	tr. 10, appr. 85%	tr. 4, appr. 85%
Lowest acc.:	tr. 9, 25%	tr. 4, 25%	tr. 1, 30%	tr. 1, 30%
In general accuracies around:	40%	70%	60%	50%
SUMMER WHEAT				
Highest acc.:	tr. 14, 80%	tr. 3, 80%	tr. 3, 85%	tr. 7, 90%
Lowest acc.:	tr. 1, 20%	tr. 3, 25%	tr. 4, 25%	tr. 1, 20%
In general accuracies around:	40%	70%	60%	50%
RYE				
Highest acc.:	tr. 3, 90%	tr.4,15, 95%	tr. 2, 95%	tr.4,14,15 100%
Lowest acc.:	tr. 5, 35%	tr. 9, 40%	tr. 5, 35%	tr. 3,9, 30%
In general accuracies around:	75%	65%	65%	60%
BARLEY				
Highest acc.:	tr. 13, 80%	tr. 15, 90%	tr. 3, 15, 95%	tr.15, 95%
Lowest acc.:	tr. 1, 10%	tr. 3, 20%	tr. 5, 40%	tr. 4, 40%
In general accuracies around:	50%	60%	50%	50%
OATS				
Highest acc.:	tr. 4,12, 70%	tr. 3,8, 90%	tr. 7, 70%	tr. 11, 65%
Lowest acc.:	tr. 1,2, 35%	tr. 10, 40%	tr. 3, 35%	tr. 7, 25%
In general accuracies around:	55%	65%	60%	50%
POTATOES				
Highest acc.:	tr. 90%	tr. 10, 85%	tr. 8, 90%	tr. 5, 70%
Lowest acc.:	tr. 4, 40%	tr. 3, 25%	tr. 1, 20%	tr. 3, 25%
In general accuracies around:	70%	60%	60%	60%
GRAZING CROPS				
Highest acc.:	tr. 95%	tr. 11, 90%	tr. 8, 90%	tr. 3,4, 95%
Lowest acc.:	tr. 5, 50%	tr. 2, 25%	tr. 10, 30%	tr. 7, 25%
In general accuracies around:	60%	50%	50%	50%

tr.= transformation

APPENDIX H

CLASSIFIED SUGAR BEET FIELDS

APPENDIX I

A DESCRIPTION OF TERMS USED IN THE TEXT

Bidirectional reflectance measurement: Spectral reflectance measurement where the angles of incidence and collection are directional.

GLAI (Green Leaf Area Index): The amount of photosynthetically active biomass is usually expressed as Green Leaf Area Index (the horizontal area of green leaves per unit area).

Hemispherical reflectance measurement: Spectral reflectance measurement where the angles of incidence and collection are hemispherical.

Irradiance: (E), is the total energy radiated onto a unit area in a unit time (watt per square metre, W/m^2).

Lambertian surface: An ideal, perfectly diffusing surface, which reflects energy equally in all directions.

Mie scattering: Multiple reflection of light waves by atmospheric particles that have the approximate dimensions of the wavelength of light. Affects long visible wavelengths.

Radiance: (L), total energy radiated by a unit area per solid angle of measurement (watt per square metre per steradian, $W/(m^2\ sr)$).

Rayleigh scattering: The wavelength dependent scattering of radiation by particles in the atmophere that are much smaller than the wavelengths scattered, affects short visible wavelengths.

Reflectance: (ρ), the ratio of the radiant energy reflected by a body to that incident upon it (unitless). Often confused with radiance. $\rho\lambda$, spectral reflectance recorded within a limited wavelength range.

Spectral irradiance: ($E\lambda$), the irradiance within a limited wavelength (watt per square metre per micrometer, $W/(m^2\ \mu m)$.

Spectral radiance: ($L\lambda$), the energy within a wavelength band, radiated by a unit area per unit solid angle of measurement (watt per square metre per steradian per micrometer , $W/(m^2\ sr\ \mu m)$).

Temporal-spectral profile: Data acquisition at only one time may be insufficient for the purpose of crop identification. The temporal pattern of spectral data represents the phenological development, the progress from seedling emergence to maturity of the crop. By relating the observed temporal-spectral pattern to the expected phenological development pattern associated with different crops, a crop identity or label can be assigned to the field.

REFERENCES

Aase, J.K., Siddoway, F.H., Millard, J.P., 1984: Spring wheat-leaf phytomass and yield estimates from airborne scanner and hand-held radiometer measurements. International Journal of Remote Sensing, 5, 771-781.

Ahlrichs, J., Bauer, M., 1983: Relation of agronomic and multispectral reflectance characteristics of spring wheat canopies. Agronomy Journal, **75**, 987-993.

Ajai, A.K., Gopland, K.S., Kamat, D.S., 1985: Discrimination of winter crops using temporal spectral profiles. International Journal of Remote Sensing, **6**, 1575-1580.

Alford, W.L., Imhoff, M.L., 1984: Radiometric accuracy of the LANDSAT-4 multispectral scanner data. In LANDSAT-4 Science Investigation Summary, NASA CP-2326, Vol. II, NASA, Washington, D.C., 69-72.

Andersson, L., Gervais, P., 1987: Marktypskartering i NV Skåne med satellit fjärranalys. Rapport 152. Swedish University of Agricultural Sciences, Department of Soil Sciences Division of Agricultural Hydrotechniques. In Swedish.

Andrieu, B., 1984: Utilisation de donnees photographiques et radiometriques pour la mise en evidence de la rhizomanie de la betterave. Proceedings of the 15th International Symposium on Remote Sensing of Environment. Ann Arbor, Michigan: Environmental Research Institute of Michigan, 813-823. In French.

Ångström, A., 1974: Sveriges klimat, 3:e ed, Generalstabens Litografiska Anstalt, 188pp. In Swedish.

Aranuvachapun, S., 1986: Satellite remote sensing of atmospheric optical depth spectrum. International Journal of Remote Sensing, **7**, 499-514.

Asrar, G., Kanemasu, E., Yoshida, M., 1985a: Estimates of leaf area index from spectral reflectance of wheat under different cultural practices and solar angle. Remote Sensing of Environment, **17**, 1-11.

Asrar, G., Kanemasu, E.T., Jackson, R.D., Pinter, P.J., 1985b: Estimation of total above-ground phytomass production using remotely sensed data. Remote Sensing of Environment, **17**, 211-220.

Badhwar, G.D., 1984a: Automatic corn-soybean classification using LANDSAT MSS data. I. Near harvest crop production estimation. Remote Sensing of Environment, **14**, 15-29.

Badhwar, G.D., 1984b: Automatic corn-soybean classification using LANDSAT MSS data. II. Early season crop proportion estimation. Remote Sensing of Environment, **14**, 31-37.

Badhwar, G.D., 1984c: Classification of corn and soybean using multitemporal thematic mapper data. Remote Sensing of Environment, **16**, 175-182.

Badhwar, G.D., 1984d: Use of LANDSAT-derived profile features for spring small-grains. International Journal of Remote Sensing, **5**, 783-797.

Badhwar, G.D., MacDonald, R.B., Metha, N.C., 1986: Satellite derived leaf area index and vegetation maps as input to global carbone cycle models - a hierarchical approach. International Journal of Remote Sensing, **7**, 265-281.

Badhwar, G.D., Garganti, C.E., Redondo, F.V., 1987: LANDSAT classification of Argentina summer crops. Remote Sensing of Environment, **2**, 111-117.

Barnett, T.L., Thompson, D.R., 1982: The use of large-area spectral data in wheat yield estimates. Remote Sensing of Environment, **12**, 509-518.

Barnett, T.L., Thompson, D.R., 1983: Large-area relation of LANDSAT MSS and NOAA-6 AVHRR spectral data to wheat yields. Remote Sensing of Environment, **13**, 277-290.

Barker, J.L., 1984: Relative radiometric calibration of LANDSAT TM reflective bands. In LANDSAT-4 Science Characterization Early Results, Vol. 3, part 2 (NASA Conference Publication, 2355pp.).

Batista, G.T., Hixon, M.M., Bauer, M.E., 1985: LANDSAT MSS crop classification performance as a function of scene characteristics. International Journal of Remote Sensing, **6**, 1521-1533.

Bauer, M.E., Daughtry, C.S., Vanderbilt, V.C., 1981: Spectral-agronomic relationships of maize, soybean and wheat canopies. International Colloquium on Spectral Signatures of Objects in Remote Sensing, 8-11 September, Avignon, France.

Best, R., Harlan, J., 1985: Spectral estimation of green leaf area index of oats. Remote Sensing of Environment, **17**, 27-36.

Bishop, Y., Feinberg, S., Holland, P., 1975: Discrete Multivariate Analysis- Theory and Practice: Cambridge Mass., The MIT Press.

Chance, J.E., Lemaster, E.W., 1985: A test of the Suits

vegetative-canopy reflectance model with LARS soybean-canopy reflectance data. International Journal of Remote Sensing, **6**, 665-672.

Charnahan, W.H., Mausel, P.W., Zhou, G.P., 1984: Evaluation of atmospheric particulate concentrations derived from analysis of ratio thematic mapper data. Machine Processing of Remotely Sensed data Symposium, 235-243.

Chhikara, R.S., 1984: Effect of mixed (boundary) pixels in crop proportion estimation. Remote Sensing of Environment, **14**, 207-218.

Cook, P., May, G., Kestle, R., 1984: A continuing development of an application for LANDSAT data: 1983 DCLC winter wheat acreage estimates for four states. Proceedings of the 18th International Symposium on Remote Sensing of Environment. Ann Arbor, Michigan: Environmental Research Institute of Michigan, 1821-1831.

Congalton, R., Mead, R., 1983: A quantitative method to test for consistency and correctness in photo-interpretation. Photogrammetric Engineering and Remote Sensing, **49**, 69-74.

Crist, E., 1984: A spectral haze diagnostic feature for normalizing LANDSAT thematic mapper data. Proceedings of the 18th International Symposium on Remote Sensing of Environment. Ann Arbor, Michigan: Environmental Research Institute of Michigan, 735-744.

Crist, E., Cicone, R., 1984a: Application of the tasseled cap concept to simulated thematic mapper data. Photogrammetric Engineering and Remote Sensing, **50**, 343-352.

Crist, E., Cicone, R., 1984b: Comparison of the dimensionality and features of simulated LANDSAT-4 MSS and TM data. Remote Sensing of Environment, **14**, 235-246.

Crist, E., 1985: A TM tasseled cap equivalent transformation for reflectance factor data. Remote Sensing of Environment, **17**, 301-306.

Crist, E.P., Kauth, R.J., 1986: The tasseled cap de-mystified. Photogrammetric Engineering and Remote Sensing, **52**, 82-86.

Csillag, F., 1986: Comparison of some classification methods on a test site (Kisköre, Hungary): Separability as a measure of accuracy. International Journal of Remote Sensing, 7, 1705-1714.

Curran, P., 1980: Relative reflectance data from preprocessed multispectral photography. International Journal of Remote Sensing, **1**, 77-84.

Curran, P., Munday, T., Milton, E., 1981: A comparison between two photographic methods for the determination of relative bidirectional reflectance. International Journal of Remote Sensing, **2**, 185-188.

Curran, P., 1982: Multispectral photographic remote sensing of green vegetation biomass and productivity. Photogrammetric Engineering and Remote Sensing, **48**, 243-250.

Curran, P., 1985: Principles of remote sensing. Longman. London. 282 pp.

Curran, P., Williamson, H.D., 1985: The accuracy of ground data used in remote sensing investigation. International Journal of Remote Sensing, **6**, 1637-1651.

Curran, P., Hay, A.M., 1986: The importance of measurement error for certain procedures in remote sensing at optical wavelengths. Photogrammetric Engineering and Remote Sensing, **52**, 229-241.

Curran, P., Williamson, H.D., 1986: Sample size for ground and remotely sensed data. Remote Sensing of Environment, **20**, 31-41.

Curran, P., Williamson, H.D., 1987a: GLAI estimation using measurements of red, near infrared, and middle infrared radiance. Photogrammetric Engineering and Remote Sensing, **53**, 181-186.

Curran, P., Williamson, H.D., 1987b: Airborne MSS data to estimate GLAI. International Journal of Remote Sensing, **8**, 57-74.

Cushnie, J., 1987: The interactive effect of spatial resolution and degree of internal variability within land-cover types on classification accuracies. International Journal of Remote Sensing, **8**, 15-29.

Dalu, G., 1986: Satellite remote sensing of atmospheric water vapour. International Journal of Remote Sensing, **7**, 1089-1097.

Dana, R.W., 1982: Background reflectance effects in LANDSAT data. Applied Optics, **21**, 4106-4111.

Dave, J.V., 1980: Effect of atmospheric conditions on remote sensing of vegetation parameters. Remote Sensing of Environment, **10**, 87-99.

Davison, G.J., 1986: Ground control pointing and geometric transformation of satellite imagery. International Journal of Remote Sensing, 7, 65-74.

Deschamps, P.Y., Herman, M., Tanre, D., 1983: Definitions of atmospheric radiance and transmittance in remote sensing. Remote Sensing of Environment, **13**, 89-92.

De Wulf, R., Goossens, R., 1986: Sugar beet biomass estimation using spectral data derived from colour infrared slides. Proceedings of the 7th International Symposium/Enchede/25-29 August 1986, Commission VII: Interpretation of Photographic and Remote Sensing Data, 239-243.

Daughtry, C.S.T., Gallo, K.P., Biehl, L.L., 1984: Spectral estimates of agronomic characteristics of crops. 1984 Machine Processing of Remotely Sensed Data Symposium, 348-356.

Duggin, M., Schoch, T., Piwinski, D., 1984: Effects of random and systematic variations in unresolved cloud on recorded radiance and on target discriminability. Applied Optics, **23**, 387-395.

Dusek, D.A., Jackson, R.D., Musick, J.T., 1985: Winter wheat vegetation indices calculated from combinations of seven spectral bands. Remote Sensing of Environment, **18**, 255-267.

Elvidge, C., Lyon, R., 1985: Influence of rock-soil spectral variation on the assessment of green biomass. Remote Sensing of Environment, **17**, 265-279.

Estes, J.E., Hajoc, E.J., Tinney, L., 1983: Fundamentals of Image Analysis: Analysis of Visible and Thermal Infrared Data. In Manual of Remote Sensing. Second edition. American Society of Photogrammetry, Falls Church, Virginia, pp.987-1124. The Sheridan Press.

European Software Contractors, 1982: UNIRAS GEOPACK GIMAGE, Interpolation and RASPACK, Users Manuals, Lyngby, Copenhagen, Denmark.

Fors, T., 1986: Skogsklassning på Ryssberget i Blekinge. Rymdbolaget. Solna. In Swedish.

Forster, B.C., 1984: Derivation of atmospheric correction procedures for LANDSAT MSS with particular reference to urban data. International Journal of Remote Sensing, **5**, 799-817.

Fusco, L., Frei, U., Travese, D., Blonda, P., Pasquariello, G., Milillo, G., 1986: LANDSAT TM image forward/reverse scan banding: characterization and correction. International Journal of Remote Sensing, **7**, 557-575.

Gallagher, J., Biscoe, P.V., 1978: Radiation absorption,

growth and yield of cereals. Journal Agric. Sci., Camb., **91**, 47-60.

Gallo, K.P., Daughtry, C.S.T., Bauer, M.E., 1985: Spectral estimates of absorbed photosynthetical active radiation in corn canopies. Remote Sensing of Environment, **17**, 221-232.

Gardner, B.R., Blad, B,L., Thompson, D.R., Henderson, K.E., 1985: Evaluation and interpretation of thematic mapper ratios in equation for estimating corn growth parameters. Remote Sensing of Environment, **18**, 225-234.

Gardner, B.R., Blad, B.L., 1986: Evaluation of spectral reflectance models to estimate corn leaf area while minimizing the influence of soil background effects. Remote Sensing of Environment, **20**, 183-193.

Goel, N., Stebel, D., 1983: Inversion of vegetation canopy reflectance models for estimating agronomic variables. I. Problem definition and initial results using the Suits model. Remote Sensing of Environment, **13**, 487-507.

Gordon, D., Philipson, W., Philpot, W., 1986: Fruit tree inventory with LANDSAT thematic mapper data. Photogrammetric Engineering and Remote Sensing, **52**, 1871-1876.

Hall-Könyves, K., 1987: The topographic effect on LANDSAT data in gently undulating terrain in southern Sweden. International Journal of Remote Sensing, **8**, 157-168.

Hall-Könyves, K., Lu Jiaju, 1987: Studies of the relationship between sugar beet parameters and digitized colour infrared aerial photography. Rapporter och Notiser No. 69. Department of Physical Geography, University of Lund. 28pp.

Harris, R., 1985: Contextual classification post-processing of LANDSAT data using a probabilistic relaxation model. International Journal of Remote Sensing, **6**, 847-866.

Hatfield, J.L., Kanemasu, E.T., Asrara, G., Jackson, R.D., Pinter, P.J., Reginato, R.J., Idso, S.B., 1985: Leaf-area estimates from spectral measurements over various planting dates of wheat. International Journal of Remote Sensing, **6**, 167-175.

Heilman, J.L., Boyd, W.E., 1986: Soil background effects on the spectral response of a three-component rangeland scene. Remote Sensing of Environment, **19**, 127-137.

Hellden, U., 1980: A test of LANDSAT-2 imagery and digital data for thematic mapping, illustrated by an environmental study in nothern Kenya. Rapporter och Notiser No. 47, Department of Physical Geography, University of Lund. 36pp.

Hinzman, L.D., Bauer, M.E., Daughtry, C.S.T., 1986: Effects of nitrogen fertilization on growth and reflectance characteristics of winter wheat. Remote Sensing of Environment, **19**, 47-61.

Hixon, M.M., Bauer, M.E., Scholz, D.K., 1982: An assessment of LANDSAT data acquisition history on identification and area estimation of corn and soybeans. Remote Sensing of Environment, **12**, 123-128.

Ho, D., Asem, A., Deschamps, P., 1986: Atmospheric correction for the sea surface temperature using NOAA-7 AVHRR and METEOSAT-2 infrared data. International Journal of Remote Sensing, **7**, 1323-1333.

Holben, B., Justice, C.O., 1980: The topographic effect on spectral response from nadir-pointing sensors. Photogrammetric Engineering and Remote Sensing, **9**, 1191-1200.

Holben, B., Justice, C.O., 1981: An examination of spectral band ratioing to reduce the topographic effect on remotely sensed data. International Journal of Remote Sensing, **2**, 115-133.

Hudson, W., Ramm, C., 1987: Correct formulation of the Kappa coefficient of agreement, Photogrammetric Engineering and Remote Sensing, **53**, 421-422.

Huete, A.R., Jackson, R.D., Post, D.F., 1985: Spectral response of a plant canopy with different soil backgrounds. Remote Sensing of Environment, **17**, 37-53.

Jackson, R.D., 1983: Spectral indices in n-space. Remote Sensing of Environment, **13**, 409-421.

Jackson, R.D., Slater, P.N., Pinter, P.J., 1983: Discrimination of growth and water stress in wheat by various vegetation indices through clear and turbid atmospheres. Remote Sensing of Environment, **13**, 187-208.

Justice, C.O., 1978: An examination of the relationship between selected ground properties and LANDSAT MSS data in an area of complex terrain in southern Italy. Proceedings of the American Society of Photogrammetry, Fall Meeting, Albuquerque, New Mexico, Falls Church, Virginia, American Society of Photogrammetry, 303-328.

Justice, C.O., Wharton, S.W., Holben, B., 1981: Application of digital terrain data to quantify and reduce the topographic effect on LANDSAT data. International Journal of Remote Sensing, **3**, 213-230.

Justice, C.O., Townshend, J., 1982: A comparison of unsupervised classification procedures on LANDSAT MSS data for an area of complex surface conditions in Basilicata, Southern Italy. Remote Sensing of Environment, **12**, 407-420.

Kamat, D.S., Gopalan, A.K.S., Ajai, Shashikumar, M.N., Sinha, S.K., Chaturvedi, G.S., Singh, A.K., 1985: Assessment of water-stress effects on crops. International Journal of Remote Sensing, **6**, 577-589.

Kaufman, Y.J., Fraser, R., 1984: Atmospheric effect on classification of finite fields. Remote Sensing of Environmnet, **15**, 95-118.

Kaufman, Y.J., 1985: Atmospheric effects on the separability of field classes measured from satellites. Remote Sensing of Environment, **18**, 21-34.

Kauth, R.J., Thomas, G.S., 1976: The tasseled cap-graphic description of spectral-temporal development of agricultural crops as seen by LANDSAT. Proceedings of the Symposium on Machine Proceedings of Remote Sensing data. 29 June - 6 July 1976. LARS, Purdue.

Kawata, Y., Haba., Y., Kusaka, T., Terashita, Y., Ueno, S., 1978: Atmospheric effects and their correction in airborne sensor and LANDSAT MSS data. Proceedings of the 12th International Symposium on Remote Sensing of Environment, Ann Arbor, Michigan: Environmental Research Institute of Michigan, 1241-1257.

Kimes, D.S., Smith, J.A., Ranson, K.J., 1980: Vegetation reflectance measurements as a function of solar zenith angle. Photogrammetric Engineering and Remote Sensing, **46**, 1563-1573.

Kimes, D.D., Markham, B., Tucker, C., McMurtey, J., 1981: Temporal relationships between spectral response and agronomic variables of a corn canopy. Remote Sensing of Environment, **11**, 401-411.

Kimes, D.S., 1983: Remote sensing of row crop structure and component temperatures using directional radiometric temperatures and inversion techniques. Remote Sensing of Environmnet, **13**, 33-55.

Kimes, D.S., Newcomb, W.W., Schutt, J.B., 1984: Directional reflectance factor distribution of a cotton row crop. International Journal of Remote Sensing, **5**, 263-277.

Kleman, J., 1985: The spectral reflectance of coniferous tree stands and barley influenced by stress. Meddelanden från Naturgeografiska Institutionen vid Stockholms Universitet, Nr A 173.

Kollenkark, J.C., Vanderbilt, V.C., Daughtry, C.S.T., Bauer, M.E., 1982: Influence of solar illumination angle on soybean canopy reflectance. Applied Optics, **21**, 1179-1184.

Kondratyev, K.Y., Kozoderov, V.V., Fedchenko, P.P., 1986: Remote sensing of the state of crops and soils. International Journal of Remote Sensing, **7**, 1213-1235.

Koo, J.Y., Kim, M., 1986: An improved hybrid classifier. International Journal of Remote Sensing, **7**, 471-476.

LANDSAT 4 DATA USERS HANDBOOK: 1984: U.S. Geological Survey, National Oceanic and Atmospheric Administraion, Sioux Falls, South Dakota.

Lennington, R.K., Sorensen, C.T., Heydorn, R.P., 1984: A mixture model approach for estimating crop areas from LANDSAT data. Remote Sensing of Environment, **14**, 197-206.

Likens, W., Wrigley, R., 1985: Impact of LANDSAT MSS sensor differences on change detection analysis. In Early Results; Volume I-Multispectral Scanner (MSS). NASA CP-2355, NASA, Washington, D.C., pp.159-175.

Lillesand, T., Kiefer, R., 1979: Remote Sensing and image interpretation. John Wiley and Sons, Inc. New York. 612pp.

Lillesand, T., Kiefer, R., 1987: Remote Sensing and image interpretation. Second edition. John Wiley and Sons, Inc. New York. 721pp.

Lindell, T., Karlsson, B., Rosengren, M., Alföldi, T., 1986: A further development of the chromaticity technique for satellite mapping of suspended sediment load. Photogrammetric Engineering and Remote Sensing, **52**, 1521-1529.

Lo, T.H.C., Scarpace, F.L., Lillesand, T.M., 1986: Use of multitemporal spectral profiles in agricultural land-cover classification. Photogrammetric Engineering and Remote Sensing, **52**, 535-544.

Loman, G., 1986: The climate of a sugar beet stand. Meddelanden från Lunds Universitets Geografiska Institution, Avhandlingar CI.

Lord, D., Desjardins, R.L., Dube, P.A., 1985a: Influence of wind on crop canopy reflectance measurements. Remote Sensing of Environment, **18**, 113-123.

Lord, D., Desjardins, R.L., Dube, P.A., Brach, E.J., 1985b: Variations of crop canopy spectral reflectance measurements under changing sky conditions. Photogrammetric Engineering and Remote Sensing, **51**, 689-695.

MacDonald, R.B., Hall, F.G., 1980: Global crop forecasting. Science, **208**, 16, 670-679.

MacFarlane, N., Robinson, I.S., 1984: Atmospheric correction of LANDSAT MSS data for multidate suspended sediment algorithm. International Journal of Remote Sensing, **5**, 561-575.

Malila, W., Metzlwe, M., Rice, D., Crist, E., 1984: Characterization of LANDSAT-4 MSS and TM digital image data. I.E.E.E. Trans. Geosci. Remote Sensing, **22**, 177.

Malila, W., 1985: Comparison of the information contents of LANDSAT TM and MSS data. Photogrammetric Engineering and Remote Sensing, **51**, 1449-1457.

Markham, B.L., Barker, J.L., 1985: Spectral characterization of the LANDSAT thematic mapper data. International Journal of Remote Sensing, **6**, 697-716.

Markham, B.K., Barker, J.L., 1986: LANDSAT MSS and TM post-calibration dynamic ranges, exoatmospheric reflectances and at-satellite temperatures. LANDSAT Technical Notes, No 1. NASA/Goddard Space Flight Center, Greenbelt, MD 20771.

Markham, B.L., Barker, J.L., 1987: Radiometric properties of U.S. processed LANDSAT MSS data. Remote Sensing of Environment, **22**, 39-71.

Mather, P.M., 1985: A commutationally-efficient maximum-likelihood classifier employing prior probabilities for remotely sensed data. International Journal of Remote Sensing, **6**, 369-376.

McCloy, K.R., Smith, F.R., Robinson, M.R., 1987: Monitoring rice areas using LANDSAT MSS data. International Journal of Remote Sensing, **8**, 741-749.

McDaniel, K.C., Hass, R.H., 1982: Assessing Mesquit-grass vegetation condition from LANDSAT. Photogrammetric Engineering and Remote Sensing, **48**, 441-450.

Megier, J., Mehl, W., 1984: Per-field classification and application to SPOT simulated, SAR and combined SAR-MSS data. Proceedings of the 18th International Symposium on Remote Sensing of Environment, Ann Arbor, Michigan: Environmental Research Institute of Michigan. 1011-1018.

Minnaert, M., 1941: The reciprocity principle in Lunar photometry. Astrophys. J., **93**, 403-410.

Moreira, M.A., Chen, S.C., Batista, G.T., 1986: Wheat-area

estimation using digital LANDSAT MSS data and aerial photographs. Internatinal Journal of Remote Sensing, **7**, 1109-1120.

Murphy, J., Butlin, T., Duff, P., Fitzgerald, A., 1984: Revised radiometric calibration technique for LANDSAT-4 thematic mapped data. IEEE Transaction on Geoscience and Remote Sensing, GE-22. No.3, May 1984.

Musick, H.B., 1986: Temporal Change of LANDSAT MSS albedo estimates in arid rangeland. Remote Sensing of Environment, **20**, 107-120.

Myers, V.I., 1983: Remote Sensing Application in Agriculture. In Manual of Remote Sensing. Second edition. American Society of Photogrammetry, Falls Church, Virginia, pp.2111-2228. The Sheridan Press.

Nelson, R.F., 1985a: Sensor-induced temporal variability of LANDSAT MSS data. Remote Sensing of Environment, **18**, 35-48.

Nelson, R.F., 1985b: Reducing LANDSAT MSS scene variability. Photogrammetric Engineering and Remote Sensing, **51**, 583-593.

Nilsson, H-E., 1985a: Remote sensing of oil seed rape infected by sclerotina stem rot and Verticillium wilt. Swedish University of Agricultural Sciences. Research Information Center, Växtskyddsrapporter. Jordbruk 33.

Nilsson, H-E. 1985b: Remote Sensing of 2-row barley infected by net blotch disease. Swedish University of Agricultural Sciences. Research Information Center, Växtskyddsrapporter. Jordbruk 34.

Nilsson, H-E. 1985c: Remote Sensing of 6-row barley infected by barley strip disease. Swedish University of Agricultural Sciences. Research Information Center, Växtskyddsrapporter. Jordbruk 36.

Odenweller, J., Johnson, K., 1984: Crop identification using LANDSAT temporal-spectral profiles. Remote Sensing of Environment, **14**, 39-54.

Olsson, L., Pinzke, S., 1984: DIAPAC software. Department of Physical Geography, University of Lund.

Parsons, A.J., 1985: Directionality in emitted infrared radiation from grassy surfaces. International Journal of Remote Sensing, **6**, 903-910.

Patel, N.K., Singh, T.P., Sahai, B., Patel, M.S., 1985: Spectral response of rice crop and its relation to yield and yield attributes. International Journal of Remote Sensing, **6**, 657-664.

Pearson, R., Miller, L., Tucker, C., 1976: A handheld spectroradiometer to estimate gramineous biomass. Applied Optics, **15**, 416-418.

Pech, R.P., Davis, A.V., Lamacraft, R.R., Graetz, R.D., 1986: Calibration of LANDSAT data for sparsely vegetated semi-arid rangelands. International Journal of Remote Sensing, **7**, 1729-1750.

Peden, D.G., Ottichilo, W.K., Mwendwa, H., 1986: Estimating pre-harvest production of maize in Kenya using large-scale aerial photography and radiometry. International Journal of Remote Sensing, **7**, 1669-1677.

Perry, C.R., Lautenschlager, L.F., 1984: Functional equivalence of spectral vegetation indices. Remote Sensing of Environment, **14**, 169-181.

Pinter, P.J., Jackson, R.D., Ezra, C.E., Gausman, H.W., 1985: Sun-angle and canopy-architecture effects on the spectral reflectance of six wheat cultivars. International Journal of Remote Sensing, **6**, 1813-1825.

Price, J.C., 1987: Calibration of satellite radiometers and the comparison of vegetation indices. Remote Sensing of Environment, **21**, 15-28.

Ranson, J., Biehl, L., Bauer, N., 1985: Variation in spectral response of soybean with respect to illumination, view and canopy geometry. International Journal of Remote Sensing, **6**, 1827-1842.

Ranson, K.J., Daughtry, C.S.T., Biehl, L., 1986: Sun angle, view angle and background effects on spectral response of simulated balsam fir canopies. Photogrammetric Engineering and Remote Sensing, **52**, 649-658.

Richardson, A.J., Wiegand, C.L., 1977: Distinguishing vegetation from soil background. Photogrammetric Engineering and Remote Sensing, **43**, 1541-1552.

Rice, D.P., Malila, W.A., 1985: Investigation of radiometric properties of LANDSAT-4 MSS. In Early Results: Volume I-Multispectral Scanner (MSS). NASA CP-2355. NASA Washington, D.C., pp.57-75.

Ripple, W., 1985a: LANDSAT thematic mapper bands for characterizing fescue grass vegetation. International Journal of Remote Sensing, **6**, 1373-1384.

Ripple, W., 1985b: Asymtotic reflectance characteristics of grass vegetation. Photogrammetric Engineering and Remote Sensing, **51**, 1915-1921.

Robinove, C., Chavez, P.S., Gehring, D., Holmgren, R., 1981: Arid land monitoring using LANDSAT albedo difference images. Remote Sensing of Environment, **11**, 133-156.

Robinove, C.J., 1982: Computation with physical values from LANDSAT digital data. Photogrammetric Engineering and Remote Sensing, 48, 781-784.

Rosenfield, G., Fitzpatric-Lins, K., 1986: A coefficient of agreement as a measure of thematic classification accuracy. Photogrammetric Engineering and Remote Sensing, **52**, 223-227.

Royer, A., Vincent, P., Bonn, F., 1985: Evaluation and correction of viewing angle effects on satellite measurements of bidirectional reflectance. Photogrammetric Engineering and Remote Sensing, **51**, 1899-1914.

Rudberg, B., 1975: Statistikens grunder. Esselte Studium. 263pp. In Swedish.

Ryan, T., Joiner, B., Ryan, F., 1981: MINITAB. Reference Manual. Pennsylvania State University.

Ryan, T., Joiner, B., Ryan, F., 1985: MINITAB Handbook. Duxbury Press. Boston. 379pp.

Ryerson, R., Dobbins, R., Thibauls, C., 1985: Timely area estimates from LANDSAT. Photogrammetric Engineering and Remote Sensing, **51**, 1735-1743.

Sabins, F.F., 1978: Remote Sensing. Principles and Interpretation. W.H. Freeman and Company. San Francisco. 426pp.

Saebö, H.V., 1985: Contextual classification of remotely sensed data: Statistical methods and development of a system. Report No. 768. Norwegian Computing Center.

Saint, G., Killmayer, A., Guyot, Riom, J., 1980: Relations between biological and phenological characteristics of vegetation and its radiometric properties: Examples in agriculture and forestry. Proceedings of the 14th International Symposium of Environment, Ann Arbor, Michigan: Environmental Research Institute of Michigan, 1069-1080.

Scarpace, F.L., Holmqvist, K.W., Fisher, L.T., 1978: LANDSAT analysis of lake quality for statewide lake classification. Proceedings of the 44th American Society of Photogrammetry, 173-195.

Shibayama, M., Wiegand, C.L., 1985: View azimuth and zenith, and solar angle effects on wheat canopy reflectance. Remote Sensing of Environment, **18**, 91-103.

Singh, S.M., Cracknell, A.P., 1986: The estimation of atmospheric effects for Spot using AVHRR channel-1 data. International Journal of Remote Sensing, **7**, 361-377.

Slater, P.N., Biggar, S.F., Holm, R.G., Jackson, R.D., Mao, Y., Moran, M.S., Palmer, J.M., Yuan, B., 1987: Reflectance and radiance-based methods for the in-flight absolute calibration of multispectral sensors. Remote Sensing of Environment, **22**, 11-37.

Smith, J.A., Tzeu, Lie Lin, Ranson, K.L., 1980: The Lambertian assumption and LANDSAT data. Photogrammetric Engineering and Remote Sensing, **9**, 1183-1189.

Sockernäringens samarbetskommitte, 1986: Plansamling för samarbetskommittens försök 1986. The Swedish Sugar Company, Staffanstorp. In Swedish.

Spanner, M., Peterson, D., Hall, M., Wrigley, R., Card, D., Running, S., 1984: Atmospheric effects on the remote sensing estimation of forest leaf area index. Proceedings of the 18th International Symposium on Remote Sensing of Environment, Ann Arbor, Michigan: Environmental Research Institute of Michigan, 1295-1303.

Steven, M.D., Biscoe, D.V., Jaggard, K.W., 1983: Estimation of sugar beet productivity from reflection in the red and infrared spectral bands. International Journal of Remote Sensing, **4**, 325-334.

Steven, M.D., Moncrieff, J.B., Mather, P.M., 1984: Atmospheric attenuation and scattering determined from multiheight multispectral scanner imagery. International Journal of Remote Sensing, **4**, 733-747.

Steven, M.D., Rollin, E.M., 1986: Estimation of atmospheric corrections for multiple aircraft imagery. International Journal of Remote Sensing, **7**, 481-497.

Stohr, C.J., West, T.R., 1985: Terrain and look angle effects upon multispectral scanner response. Photogrammetric Engineering and Remote Sensing, **51**, 229-235.

Story, M., 1986: Accuracy assessment: A user's perspective. Photogrammetric Engineering and Remote Sensing, **52**, 397-399.

Strahler, A., Logan, T.L., Bryant, N.A., 1978: Improving forest cover classification accuracy from LANDSAT by incorporating topographic information. Proceedings of the 12th International Symposium on Remote Sensing of Environment, Vol. 2, Ann Arbor, Michigan: Environmental Research Institute of Michigan, 927-942.

Suits, G., 1983: Extension of a uniform canopy reflectance model to include row effects. Remote Sensing of Environment, **13**, 113-129.

Suits, G., 1985: An analysis of spectral discrimination between crop and soybean using a row crop reflectance model. Remote Sensing of Environment, **17**, 109-116.

Swain, P.H., Davis, S.M., 1978: Remote Sensing: The Quantitative Approach. McGraw Hill. New York. 395pp.

Teillet, P.M., 1986: Image Correction for radiometric effects in remote sensing. International Journal of Remote Sensing, **7**, 1637-1651.

Tilton, J., Markham, B., Alford, W., 1985: LANDSAT-4 and LANDSAT-5 MSS Coherent Noise: Characterization and Removal. Photogrammetric Engineering and Remote Sensing, **51**, 1263-1279.

Townshend, J., 1984: Agricultural land-cover discrimination using thematic mapped spectral bands. International Journal of Remote Sensing, **5**, 681-698.

Troedsson, T., Wiberg, M., 1986: Soil Map of Sweden. Swedish University of Agricultural Sciences. Department of Forest Soils. In distribution, The Royal Swedish Academy of Agriculture and Forestry.

Tucker, C.J., Vanpraet, C.K., Sharman, M.J., Van Itterman, C., 1985: Satellite remote sensing of total herbaceous biomass production in the Senegalese Sahel: 1980-1984. Remote Sensing of Environment, **17**, 233-249.

Tucker, C.J., Sellers, P.J., 1986: Satellite remote sensing of primary production. International Journal of Remote Sensing, **7**, 1395-1416.

Turner, R., Spencer, M., 1972: Atmospheric model for correction of spacecraft data. Proceedings of the 8th International Symposium on Remote Sensing of Environment, Ann Arbor, Michigan: Environmental Research Institute of Michigan, 895-934.

Turner, R., 1975: Signature variations due to atmospheric effects. Proceedings of the 10th International Symposium on Remote Sensing of Environment, Ann Arbor, Michigan: Environmental Research Institute of Michigan, 671-694.

Wardley, N.W., Curran, P.J., 1984: The estimation of green-leaf-area index from remotely sensed airborne multispectral data. International Journal of Remote Sensing, **5**, 671-679.

Wastenson, L., 1978: Kartläggning av grödor med multispektral svepradiometri från flygplan. Ett metodförsök vid Ultuna, Uppland. Naturgeografiska institutionen. Stockholms universitet. Forskningsrapport 33. In Swedish.

Wiegand, C.L., Richardson, A.J., 1982: Comparisons among a new soil index and other two-and four-dimensional vegetation indices. Proceedings of the 48th Annual Meeting. American Society of Photogrammetry, 211-227.

Weiser, R.L., Asrar, G., Miller, G.P., Kanemasu, E.T., 1986: Assessing grassland biophysical characteristics from spectral measurements. Remote Sensing of Environment, **20**, 141-152.

Williams, R.B.G., 1984: Introduction to Statistics for Geographers and Earth Scientists. Macmillan Publishers LTD. 349pp.

Yool, S., Star, J., Estes, J., Botkin, D., Eckhard, D., Davis, F., 1986: Performance analysis of image processing algorithms for classification of natural vegetation in the mountains of Southern California. International Journal of Remote Sensing, 7, 683-702.

MEDDELANDEN FRÅN LUNDS UNIVERSITETS GEOGRAFISKA INSTITUTION

Serie Avhandlingar

I. *Herman Richter:* Skånes karta från mitten av 1500-talet till omkring 1700. Bidrag till en historisk-kartografisk undersökning. (1929).

II. *Josef Westin:* Kulturgeografiska studier inom Nätra-, Näske- och Utbyåarnas flodområden samt angränsande kusttrakter. (1930).

III. *Herman Richter–Wilhelm Norlind:* Orbis Arctoi Nova et Accurata Delineatio Auctore Andrea Bureo Sueco 1626. (1936).

IV. *Sven Björnsson:* Sommen-Åsundenområdet. En geomorfologisk studie. (1937).

V. *Arne Sandell:* Tektonik och morfologi inom dalformationen med omgivande urbergsterräng. (1941).

VI. *Sven Dahl:* Torna och Bara. Studier i Skånes bebyggelse- och näringsgeografi före 1860. (1942).

VII. *Karl Erik Bergsten:* Isälvsfält kring norra Vättern. Fysisk-geografiska studier. (1943).

VIII. *Carl Erik Nordenskjöld:* Morfologiska studier inom övergångsområdet mellan Kalmarslätten och Tjust. (1944).

IX. *Sven Björnsson:* Blekinge. En studie av det blekingska kulturlandskapet. (1946).

X. *Karl Erik Bergsten:* Östergötlands bergslag. En geografisk studie. (1946).

XI. *Tor Holmquist:* Den halländska vinterfiskehamnsfrågan. (1947).

XII. *Olof Ängeby:* Landformerna i nordvästra Jämtland och angränsande delar av Nord-Tröndelag. (1947).

XIII. *Axel Wennberg:* Lantbebyggelsen i nordöstra Östergötland 1600-1875. (1947).

XIV. *Lars Bjerning:* Skånes jord- och stenindustri. Dess utveckling, lokalisering och betydelse ur näringsgeografisk synvinkel. (1947).

XV. *Allan Weinhagen:* Norbergs bergslag samt Gunnilbo och Ramnäs till omkring 1820. Studier i områdets närings- och bebyggelsegeografi. (1947).

XVI. *Helge Stålberg:* Smålands skogs- och träförädlingsindustrier. En näringsgeografisk studie. (1947).

XVII. *Folke Lägnert:* Veteodlingen i södra och mellersta Sverige. (1949).

XVIII. *Yngve Nilsson:* Bygd och näringsliv i norra Värmland. En kulturgeografisk studie. (1950).

XIX. *Olof Ängeby:* Evorsionen i recenta vattenfall. (1951).

XX. *Karl Erik Bergsten:* Sydsvenska födelseortsfält. (1951).

XXI. *Folke Lägnert:* Valmanskåren på Skånes landsbygd 1911-1948. (1952).

XXII. *Olof Nordström:* Relationer mellan bruk och omland i östra Småland 1750-1900. (1952).

XXIII. *Arvid Bergdahl:* Israndsbildningar i östra Syd- och Mellansverige med särskild hänsyn till åsarna. (1953).

XXIV. *Sven E Behrens:* Morfometriska, morfogenetiska och tektoniska studier av de nordvästskånska urbergsåsarna, särskilt Kullaberg. (1953).

XXV. *Torsten Hägerstrand:* Innovationsförloppet ur korologisk synpunkt. (1953).

XXVI. *Gunhild Weimarck:* Studier över landskapets förändring inom Lönsboda, Örkeneds socken, nordöstra Skåne. (1953).

XXVII. *Ingemar Larsson:* Structure and Landscape in Western Blekinge, Southeast Sweden. (1954).

XXVIII. *Sven Godlund:* Busstrafikens framväxt och funktion i de urbana influensfälten. (1954).

XXIX. *Folke Lägnert:* Syd- och mellansvenska växtföljder. Del I. De äldre brukningssystemens upplösning under 1800-talet. (1955).

XXX. *Olof Ängeby:* Toppkonstans, erosionsytor och passdalar i Jämtland och Tröndelag. (1955).

XXXI. *Gunnar Johnsson:* Glacialmorfologiska studier i södra Sverige. (1956).

XXXII. *Folke Lägnert:* Syd- och mellansvenska växtföljder. Del II. 1900-talet. (1956).

XXXIII. *Olof Nordström:* Befolkningsutveckling och arbetskraftsproblem i östra Småland 1800-1955. (1957).

XXXIV. *Sven Godlund:* Befolkning–regionsjukhus–resmöjligheter–regioner. (1958).

XXXV. *Elis Pålsson:* Gymnasiers rekrytering och lokalisering. (1958).

XXXVI. *Harald Svensson:* Glaciation och morfologi. En glacialgeografisk studie i ett tvärsnitt genom Skanderna mellan södra Helgelandskusten och Kultsjödalen. (1959).

XXXVII. *Erik Ljungner:* Nahuel Huapi. Ein geographischer Querschnitt durch die Anden in Patagonien. (1959).

XXXVIII. *Nils Lewan:* Om pendling mellan bostad och arbetsplats. En undersökning med material från sydvästra Skåne. (1960).

XXXIX. *Åke Mattsson:* Morphologische Studien in Südschweden und auf Bornholm über die nichtglaziale Formenwelt der Felsenskulptur. (1962).

XL. *Stig Nordbeck:* Framställning av kartor med hjälp av siffermaskiner. (1964).

XLI. *Olof Nordström:* Svensk glasindustri 1550-1960. (1962).

XLII. *Jan Davidsson:* Littoral Processes and Morphology on Scandian Flatcoasts. (1963).

XLIII. *Martin Markgren:* Detaljmorfologiska studier i fast berg och blockmaterial. Geomorfologisk studie inom Fennoskandia med Skåne. (1962-1963).

XLIV. *Martin Markgren:* Geomorphological Studies in Fennoscandia. II. Chute Slopes in Northern Fennoscandia. A. Regional Studies. 1964).

XLV. *Martin Markgren:* Geomorphological Studies in Fennoscandia. II. Chute Slopes in Northern Fennoscandia. B. Systematic Studies. (1964).

XLVI. *Lennart Améen:* Stadsbebyggelse och domänstruktur. (1964).

XLVII. *Arvid Bergdahl:* Det glaciala landskapet. (1961).

XLVIII. *Olof Nordström–Solveig Mårtensson:* Turism på Öland. (1966).
XLIX. *Jan O Mattsson:* The Temperature Climate of Potato Crops. (1966).
L. *Nils Lewan:* Landsbebyggelse i förvandling. (1967).
LI. *Gösta Nordholm:* Skånes äldre ekonomiska geografi. (1967).
LII. *Sven Godlund–Torsten Hägerstrand–Bengt Svanström:* Samhällsutvecklingen och samhällsplaneringen. (1967).
LIII. *Tor Fr Rasmussen:* Storbyutvikling og arbeidsreiser. (1966).
LIV. *Erik Fagerlund–Harald Svensson–Sven Lindqvist m fl:* Infrarödtermografi. Principer och naturgeografiska tillämpningar. (1967).
LV. *Lars Eldblom:* Structure foncière, organisation et structure sociale. (1968).
LVI. *Knut Norborg:* Jordbruksbefolkningen i Sverige. (1968).
LVII. *Gunhild Weimarck:* Ulfshult (1968). Out of print.
LVIII. *Rune Järvenstedt–Sven Lindqvist–Jan O Mattsson m fl:* Televisionssystem i naturgeografisk forskning. (1968).
LIX. *Arne Jakobsson:* Omflyttningen i Sverige 1950-1960. (1969).
LX. *Åke Hillefors:* Västsveriges glaciala historia och morfologi. (1969). Out of print.
LXI. *Sven Lindqvist:* Bebyggelseklimatiska studier. (1970).
LXII. *Torsten Hägerstrand–Gunnar Törnqvist m fl:* Urbaniseringen i Sverige. (SOU 1970:14).
LXIII. *Bengt Sahlberg:* Interregionala kontaktmönster. Personkontakter inom svenskt näringsliv – en flygpassagerarstudie. (1970).
LXIV. *Björn Hedberg:* Kontaktsystem inom svenskt näringsliv. En studie av organisationers externa personkontakter. (1970).
LXV. *Mats G Engström:* Regional arbetsfördelning. Nya drag i förvärvsarbetets geografiska organisation i Sverige. (1970).
LXVI. *Torsten Persson:* Geomorphological Studies in the South-Swedish Highlands. (1972).
LXVII. *Dewitt Davis Jr.:* A factoral ecology of Malmö 1965. The social geography of a city. (1972). Out of print.
LXVIII. *Zoltan Petery:* Studier i svensk regionalplanering. Regionalplanering enligt byggnadslagen i mindre regioner. (1972). Out of print.
LXIX. *Tommy Book:* Stadsplan och järnväg i Norden. (1974).
LXX. *Hans Abrahamson:* Studier i jordbrukets omstrukturering. (1974).
LXXI. *Christer Persson:* Kontaktarbete och framtida lokaliseringsförändringar. (1974).
LXXII. *Ulf Helldén:* Karst. En studie av Artfjällets karstområde samt jämförande korrosionsanalyser från Västspetsbergen och Tjeckoslovakien. (1974).
LXXIII. *Jànos Szegö:* Befolkningstäthet, markanvändning, planering (vol 1 o 2). (1974).
LXXIV. *Raul Nino Guerrero:* Rural to Urban Drift of the Unemployed in Colombia. (1975).
LXXV. *Ulf Erlandsson:* Företagsutveckling och utrymmesbehov. (1975).
LXXVI. *Sture Öberg:* Methods of Describing Physical Access to Supply Points. (1976).
LXXVII. *Bo Lenntorp:* Paths in Space-time Environments. A Time-geographic Study of Movement Possibilities of Individuals. (1976).
LXXVIII. *Richard Åhman:* Palsar i Nordnorge. En studie av palsars morfologi, utbredning och klimatiska förutsättningar i Finnmarks och Troms fylke. (1977).

LXXIX. *Björn Gyllström:* The Organization of Production as a Space-modelling Mechanism in Underdeveloped Countries. (1977).

LXXX. *Anders Järnegren–Fosco Ventura:* Tre samhällens förändringshistoria. Exploatering av den fysiska miljön i historisk belysning. (1977).

LXXXI. *Tommy Book:* Stadsplan och järnväg i Storbritannien och Tyskland. (1978).

LXXXII. *Jan O Mattsson–Leif Börjesson:* Lokalklimatiska temperaturstudier inom ett skånskt fruktodlingsdistrikt med särskilt beaktande av frostläntheten. (1978).

LXXXIII. *Bjørn Terje Asheim:* Regionale ulikheter i levekår. (1979).

LXXXIV. *Solveig Mårtensson:* On the Formation of Biographies in Space-time Environments. (1979).

LXXXV. *Erik Wallin:* Vardagslivets generativa grammatik – Vid gränsen mellan natur och kultur. (1980).

LXXXVI. *Reinhold Castensson:* Välja för framtid – om markanvändningsval och förtroendemannainflytande i kommunal planering. (1980).

LXXXVII. *Kerstin Montal:* Industri och vatten. Den vattenförorenande industrins lokaliseringsproblem i Malmöhus län. (1980).

LXXXVIII. *Tommy Carlstein:* Time Resources, Society and Ecology. (1980).

LXXXIX. *Jonas Åkerman:* Studies on periglacial geomorphology in West Spitsbergen. (1980).

XC. *Leif Engh:* Karstområdet vid Lummelunds bruk, Gotland, med speciell hänsyn till Lummelundagrottan. (1980).

XCI. *Karna Lidmar-Bergström:* Pre-quaternary Geomorphological Evolution in Southern Fennoscandia. (1982).

XCII. *Lars-Olof Olander:* Staten, kommunerna och servicen. (1984).

XCIII. *Bo Malmström–Owe Palmér:* Glacial och periglacial geomorfologi på Varangerhalvön, Nordnorge. Geomorfologisk kartering med analys av glaciala former och blockhav. (1984).

XCIV. *Franz-Michael Rundquist:* Hybrid Maize Diffusion in Kenya. (1984).

XCV. *Girma Yadeta:* Dynamic Processes of Development in Marginal Areas. (1985).

XCVI. *Anders Sporrek:* Food Marketing and Urban Growth in Dar Es Salaam. (1985).

XCVII. *Rolf Nyberg:* Debris Flows and Slush Avalanches in Northern Swedish Lappland. Distribution and Geomorphological Significance. (1985).

XCVIII. *Lennart Olsson:* An Integrated Study of Desertification – Applications of Remote Sensing, GIS and Spatial Models in Semi-arid Sudan. (1985).

XCIX. *Mikael Stern:* Census from Heaven? (1985).

C. *Katarina Olsson:* Remote Sensing for Fuelwood Resources and Land Degradation Studies in Kordofan, the Sudan. (1985).

CI. *Göran Loman:* The Climate of a Sugar Beet Stand – Dynamics, Impact on the Crop and Possibilities of Improvement. (1986).

CI. *Eric Clark:* The Rent Gap and Urban Change. Case Studies in Malmö 1860-1985. (1987).

CII. *Karin Hall-Könyves:* Remote Sensing of Cultivated Lands in the South of Sweden. (1988).

CIII. *Eva Ahlcrona:* The Impact of Climate and Man on Land Transformation in Central Sudan. Applications of Remote Sensing. (1988).